GLOSSARY ON SOLID WASTE

GLOSSARY ON SOLID WASTE

Compiled by
P.K. Patrick
Waste Management Consultant

WORLD HEALTH ORGANIZATION
Regional Office for Europe
COPENHAGEN
1980

ISBN 92 9020 199 1

© World Health Organization 1980

Publications of the WHO Regional Office for Europe enjoy copyright protection in accordance with the provisions of Protocol 2 of the Universal Copyright Convention. For rights of reproduction or translation, in part or *in toto*, of this publication application should be made to the WHO Regional Office for Europe, Scherfigsvej 8, DK-2100 Copenhagen Ø, Denmark. The Regional Office welcomes such applications.

The designations employed and the presentation of the material in this publication do not imply the expression of any opinion whatsoever on the part of the Secretariat of the World Health Organization concerning the legal status of any country, territory, city or area or of its authorities, or concerning the delimitation of its frontiers or boundaries.

The mention of specific companies or of certain manufacturers' products does not imply that they are endorsed or recommended by the World Health Organization in preference to others of a similar nature that are not mentioned. Errors and omissions excepted, the names of proprietary products are distinguished by initial capital letters.

PRINTED IN DENMARK

REVIEWERS

Mr H. Brown, Sumner and Partners Consulting Engineers, 251 High Street, Orpington, Kent BR6 0PH, United Kingdom

Dr J.A. da Costa, Division of Water Sciences, UNESCO, 7 Place de Fontenoy, 75007 Paris, France

Professor D.J. Hagerty, Department of Civil Engineering, University of Louisville, KY 40208, USA

Mr H. Lanier Hickman, Jr., Director of Operations for Solid Wastes Management Program, US Environmental Protection Agency, Washington, DC 20460, USA

Mr P.W. Jolly, Sanitarian, WHO Project TUR/BSM 001, Government Office for WHO, P.K. 235, Yenisehir, Ankara, Turkey

Mr J. Marriott, Engineer, Surveyor and Public Health Inspector, The Clefte, No. 1 Wharfe View, Wetherby, Yorks LS22 4HB, United Kingdom

Mr J. Skitt, County Waste Disposal Engineer, Staffordshire County Council, 4 Chapel Street, Stafford ST16 2BX, United Kingdom

Mr E. Zehnder, Chemical Engineering Consultant, Nufenstrasse 34, 4054 Basel, Switzerland

INTRODUCTION

The technology and methods of solid waste management are very much in the evolutionary stage, but form essential elements of environmental pollution control. Solid waste management is interdisciplinary, involving many branches of engineering, chemistry, hydrogeology and other sciences, but while much of the terminology used in the literature on solid waste management emanates from these disciplines, a considerable amount of specialized terminology has evolved.

Many of the specialized terms may not be readily understandable to laymen and administrators not familiar with the subject. Confusion can also arise because different terms may be used to describe the same activity or equipment; this is particularly the case in the English language, where differences in terminology have developed, for example, between the United Kingdom and the USA (e.g., "controlled tip" and "sanitary landfill" mean the same thing). Where terms have a usage peculiar to one country, this has been indicated in parentheses, e.g., (USA usage).

The WHO Regional Office for Europe, whose environmental health programme includes a solid waste management component, considered that there was a need to produce a glossary to reduce possibilities of misunderstanding, and facilitate communication in this increasingly important field.

The preparation of a glossary inevitably presents many difficulties in attempting to unify terms and concepts. Definitions were taken from existing glossaries and other similar works of reference. The source of these definitions is given in brackets. Where no suitable definitions could be found, they were drafted by the compiler. It should be emphasized that publication of the terms and definitions in this glossary does not imply that they are recommended by WHO.

The WHO Regional Office for Europe would like to express its indebtedness to all those who have contributed to the glossary, particularly the reviewers listed on p. v, who read and commented on the draft, and Mr D.A. Lowe, Chief, Technical Terminology Service, WHO, Geneva, who gave invaluable assistance in finalizing the glossary. The WHO Regional Office for Europe thanks all the organizations and publishers who have given permission to use definitions from their publications.

This glossary is published in the hope that it will prove useful to all persons involved professionally or otherwise in solid waste management. This edition is to be regarded as provisional. Suggestions from users for modifications, corrections or additions in a future revised edition are welcomed and should be addressed to: *Director, Promotion of Environmental Health, WHO Regional Office for Europe, Scherfigsvej 8, 2100 Copenhagen Ø, Denmark.*

HOW TO USE THE GLOSSARY

The terms are listed in alphabetical order on a letter-by-letter basis, and punctuation has been ignored. In general, compound terms, comprising a noun and an adjective, have been inverted and placed under the noun (e.g., drain, agricultural); however, where the adjective is the really significant part of the term and the noun very general in nature (e.g., organic content), or in the case of certain compound terms that do not consist of a noun and an adjective (e.g., night soil), such terms have not been inverted. Where a number of different types of a given machine, process, etc., are defined, these have been grouped under a single entry.

A

abrasion Wearing away of surface material, such as refractories in an incinerator or parts of solid waste handling equipment, by the scouring action of moving solids, liquids, or gases. (EPA)

acceptance test A test of plant or equipment supplied under contract to determine whether performance is in accordance with specification. See also *commissioning test*

accumulator, hydraulic A chamber fitted to the delivery side of a reciprocating pump for the storage of energy.

acre-foot A non-SI unit of volume used in some countries, defined as the volume of material that would cover an area of 1 acre to a depth of 1 foot. The SI unit is the cubic metre: 1 acre-foot = 1233.5 m^3.

actinomycetes A large group of mould-like microorganisms that play a significant part in the stabilization of solid waste by composting.

adsorption A process whereby one or more components of an interfacial layer between two bulk phases are either enriched or depleted. If the process is one of enrichment, it is known as positive adsorption or simply adsorption; if it is one of depletion, it is referred to as negative adsorption. The forces involved may be either chemical (valence forces), in which case the process is termed *chemisorption* (or chemical adsorption), or intermolecular, in which case the term *physisorption* (or physical adsorption) is used. "Adsorption" should be carefully distinguished from "absorption".

aeration The process of exposing a bulk material, such as compost, or solid waste in a landfill, to intimate contact with air, or of charging a liquid with air. Several methods may be used, e.g., forcing air into the material or through the liquid, or agitating the liquid to promote surface adsorption of air.

aerobe, facultative An organism that can respire in either the presence or the absence of atmospheric oxygen.

aerobe, obligate An organism that requires free (atmospheric) oxygen for respiration.

after-burner A combustion chamber in which any smoke, carbon monoxide, or other organic substances in the exhaust gases from furnaces or certain types of engine can be completely burned. The chamber normally contains oil- or gas-fired burners to maintain the necessary high temperature, and the oxygen content of the gases is adjusted by admitting air as required. After-burners are also used to burn malodorous organic substances in gases arising from various processes. If the gases entering the after-burner are sufficiently hot, the use of a catalyst sometimes enables complete combustion to be achieved at a lower than normal temperature, and the provision of fuel to the after-burner may be necessary only when starting up or at low throughput. Catalytic after-burners are widely used in industry and are often fitted to certain types of waste incinerator, e.g., small on-site domestic incinerators and specialized types of industrial waste incinerator.

after-use The intended or planned use of land restored by tipping of solid waste, e.g., as agricultural land or parkland.

aggregate 1. In gas cleaning, a relatively stable assembly of dry particles, formed under the influence of physical forces. (Provisional ISO, 2) The process of aggregate formation is known as *aggregation.* In other contexts, the particles need not be dry. The term is used in colloid chemistry to refer specifically to the structure that results from the cohesion of colloidal particles. Aggregate formation in colloidally unstable sols is termed either *coagulation* (formation of a coagulum) or *flocculation* (formation of a floc); some authors make a distinction between the two terms, but in more general usage they are synonymous. 2. Crushed rock or gravel screened to specific sizes for use in road surfaces, concrete, or bituminous mixes. Incinerator clinker may be used as an aggregate for some purposes.

aggregation See *aggregate*

air deficiency A lack of air, in an air–fuel mixture, to supply the quantity of oxygen stoichiometrically required to completely oxidize the fuel. (EPA)

air jet [One of several] streams of high-velocity air that issue from nozzles in an incinerator enclosure to provide turbulence, combustion air, or a cooling effect. (EPA)

air, secondary See *overgrate air*

air, theoretical The amount of air, calculated from the chemical composition of a waste, that is required to completely burn the waste. Also referred to as stoichiometic air and theoretical combustion air. (EPA)

amplitude The maximum value of a sinusoidal quantity. (ISO, 1) An example of the use of the term in solid waste applications is the amplitude of the motion of a vibratory screen or feeder.

analysis (solid waste) The determination of the types and proportions of the various materials comprising a given sample of solid waste. Recommended procedures for carrying out some solid waste analyses are available, e.g., those of the Institute of Solid-Wastes Management in the United Kingdom.

analysis, elemental The determination of the elements in a sample, with no attempt to discover the way in which the elements are combined to form molecules.

analysis, proximate The determination of the classes or groups of related components in a sample, with no attempt to identify or determine each individual component (e.g., the determination of total alkaloids in a plant, or the determination of total protein in blood serum). Proximate analysis is frequently applied to solid fuels in order to determine their moisture, ash and fixed carbon contents; certain other components may also be determined, together with the calorific value of the fuel.

analysis, ultimate See *analysis, elemental*

angle of repose The maximum acute angle that the inclined surface of a pile of loosely divided material can make with the horizontal. (EPA)

animal bedding Material, usually organic, placed on the floor of stables, cowsheds, etc. to absorb excreta.

aquifer Porous water-bearing formation (bed or stratum) of permeable rock, sand, or gravel capable of yielding significant quantities of water. (WMO/UNESCO)

ash The solid residue of effectively complete combustion. (Provisional ISO, 2)

ash-free basis The method whereby the weight of ash in a fuel sample is subtracted from its total weight and the adjusted weight is used to calculate the percent of certain constituents present. For example, the percent of fixed carbon (FC) on an ash-free basis is computed as follows:

$$\frac{FC\,(weight) \times 100}{fuel\,sample\,(weight) - ash\,(weight)} = \%\ \text{ash-free FC} \quad \text{(EPA)}$$

ash-handling system, pneumatic A system of pipes and cyclone separators that conveys fly ash or floor dust to a bin via an air stream. (EPA)

ashpit 1. A fixed chamber for the storage of solid waste. Formerly common in domestic premises but now obsolete and replaced by movable containers. 2. A pit located below a furnace in which ash is accumulated and from which it is removable.

ash sluice A trench or channel in which water transports residue from an ash pit to a disposal or collection point. (EPA)

aspect ratio The ratio of length to width, e.g., of an incinerator grate.

attemperation The process of cooling combustion gases. *Air a.*, the cooling of combustion gases by mixing air with them. *Water a.*, a method of cooling combustion gases by evaporating water mixed with the gases in the form of a fine spray.

attenuation The natural purification of polluted water by passage through the ground; the gradual removal of suspended materials from a liquid by passage through porous material.

autoignition point See *ignition point* (2)

autothermic Capable of sustaining combustion without an external source of heat.

availability factor The period of time during which equipment or plant is available for use, expressed as a percentage of nominal total available time; thus, for example:

$$\text{Annual availability factor} = \frac{\text{(No. of days on which plant is capable of being used)}}{365} \times 100$$

B

backactor See *backhoe*

back-blading A technique used in levelling solid waste or spreading covering material in landfill operations. The tractor draws the cutting edge of the blade backwards over the waste or covering material.

backfill The process of refilling an excavation, or the material used for this purpose.

backhoe A mechanized trench-digging tool used with a hydraulic excavator; it is used in some transfer stations to compact solid waste in top-loading transfer vehicles.

backhoe tamping A process used in some refuse transfer stations in which a conventional backhoe is used to compact solid waste in open-top transfer vehicles.

baffle A device intended to change the direction or reduce the velocity of a fluid.

baffle chamber A type of settling chamber, containing a system of baffles, in which coarse particulate matter (e.g., fly ash) is removed from stack gases by alteration of its direction or reduction of its velocity.

bag filter See *filter bag*

bagging A process in which matured dried compost is put into lined or impervious bags or socks for storage and sale.

bale To form into a dense mass by the application of high pressure in a baling machine. Waste paper and scrap metal are commonly baled to facilitate handling, storage and transportation.

baler A machine used to form, by compression, a bale of solid waste or other material.

barge bed A mud bottom alongside a jetty or the bank of a river where barges can moor and sit on the mud at low tide.

barrier loader A type of refuse collection vehicle in which refuse is loaded over a vertical plate or barrier, initially placed close to the front of the body and moved towards the rear as loading progresses.

batch process A process in which raw materials are fed into a plant in discrete batches rather than continuously.

bearing capacity The maximum load that a material can support before failing. (EPA)

beater arms Blades attached to a contrarotating shaft in some drum-type pulverizers in order to assist the pulverization process.

Beccari® process A refuse composting process in which the refuse is enclosed in ventilated cells for 40–50 days, after which decomposed material is separated from the noncompostible elements.

bentonite A clay mineral similar to fuller's earth; it has been used as an impermeable lining material in waterlogged ground, e.g., in the construction of refuse storage bunkers, and as a lining for mineral excavations to be used for the tipping of solid waste, with the object of preventing polluting leachate from reaching groundwater.

bin A portable receptacle fitted with a lid, for the storage of solid waste at domestic or commercial premises.

binette A small portable refuse container with a closely fitting lid, for use within the dwelling, from which refuse is emptied into containers to await collection. (BSI)

bin hoist mechanism The device fitted to a refuse collection vehicle whereby bins or containers are lifted and discharged.

bin liner A plastic bag or paper sack that fits freely into a bin and is removed for collection together with its contents.

bin trolley A light two-wheeled trolley used for moving dustbins, and particularly those of the heavy steel "dustless-loading" type, to the collection vehicle.

biochemical oxygen demand (BOD) Index of water pollution which represents the content of biochemically degradable substances in the water. (WMO/UNESCO)

biodegradable Capable of being broken down physically and/or chemically by the action of microorganisms.

blast gate A sliding metal damper in a duct, generally used to regulate the flow of forced air or to allow its escape at abnormally high pressure.

block, insulating See *brick, fireclay*

blow hole A hole or gap formed in the layer of material burning on an incinerator grate, caused by flash burning of random concentrations of highly volatile material, or by uneven burning of unequal thicknesses of the firebed.

BOD See *biochemical oxygen demand*

boiler feed-water Water used for steam-raising and usually requiring special treatment.

boiler, integral A boiler enclosed within the structure of an incinerator and fired directly by combustion gases.

boiler rating The heating capacity of a boiler expressed in appropriate units.

boiler, waste-heat A boiler fired by gases rejected from a process; it is normally a separate unit, set apart from the main furnace structure.

boning rod A T-shaped board used for sighting and lining up when setting out levels.

boom Any heavy beam that is hinged at one end and carries a weight-lifting device at the other. (EPA)

booster cycle The period during which additional hydraulic pressure is exerted to push the last charge of solid waste into a transfer trailer or a container attached to a stationary compactor. (EPA)

borehole A hole drilled into the ground in order to obtain samples of the underlying strata. In USA usage, also known as a *sampling hole*.

breaker bar Part of a stationary compactor serving to break bulky items passing from the charging chamber to the compaction chamber. (POLLOCK)

breeching A passage that conveys the products of combustion to a stack or chimney. (EPA)

breeching bypass An arrangement whereby breechings and dampers permit the intermittent use of two or more passages to direct or divert the flow of the products of combustion. (EPA)

brick, fireclay (or **firebrick**) Refractory brick made from fireclay. *Alumina-diaspore f.b.*, brick consisting mainly of diaspore or nodule clay and having an alumina content of 50, 60, or 70 percent (plus or minus 2.5 percent). *High-duty f.b.*, a fireclay brick that has a pyrometric cone equivalent (PCE) not lower than Cone 31-23, or does not deform more than 1.5 percent at 2460 °F (1350 °C) in the standard local test. *Insulating f.b.*, a firebrick having a low thermal conductivity and a bulk density of less than 70 pounds per cubic foot [1 121 kg/m^3]; suitable for lining industrial furnaces. Also called *insulating block*. *Intermediate-duty f.b.*, a fireclay brick that has a PCE above Cone 29 or does not deform more than 3 percent at 2460 °F (1350 °C) in the standard local test. *Super-duty f.b.*, a fireclay brick that has a PCE above Cone 33 on the fired product, shrinks less than 1 percent in the American Society for Testing Materials permanent linear change test, Schedule C (2910 °F [1 598 °C]), and does not incur more than 4 percent loss in the panel spalling test (preheated to 3 000 °F [1 649 °C]). (All EPA)

brick, insulating See *brick, fireclay*

bridging The wedging of solid waste across a hopper or chute.

briquetter A machine that compresses a material, such as metal turnings or coal dust, into small pellets. (EPA)

British thermal unit A non-SI unit of work or energy, defined as the quantity of heat required to raise the temperature of 1 lb of water by 1 °F or, more precisely, from 39 °F to 40 °F. Factor for converting British thermal units (Btu) into joules (J): 1 Btu $\approx$ 1.055 06 $\times$ 10^3 J.

bubble washer See *dust separator*

bucket An open container affixed to the movable arms of a wheeled or tracked vehicle to spread solid waste and cover material, and to excavate soil. (EPA)

bucket elevator One (or two linked) endless chains with buckets attached for raising loose material such as slurry, coal, or stone at slopes varying from 45° to nearly vertical. (SCOTT)

bucket loader A wheeled or tracked vehicle fitted with a bucket container on movable arms and used to excavate soil or cover material at a controlled tip, to load vehicles, or to stockpile material.

buckstays Metal sections with which furnace brickwork is braced externally to maintain rigidity.

buffer storage Storage space provided at a solid waste treatment plant or transfer station to allow collection vehicles to discharge their loads without delay and without affecting the rate of processing or transfer of the solid waste. Buffer storage may take the form of platform space or a bunker or hopper. At a transfer station, trailers or containers may act as buffer storage.

bulk load vehicle See *transfer vehicle*

bulking of sludge The process whereby an activated sludge settles poorly because of a low-density floc.

bull clam A tracked vehicle that has a hinged, curved bowl on the top of the front of the blade. (EPA)

bund 1. A barrier of inert material constructed across a site containing water so as to form a lagoon. 2. Walls forming an enclosure round a tank so as to contain the contents in the event of leakage.

bunker A concrete or steel pit for the temporary storage of solid waste.

burn (brick) The degree of heat treatment to which refractory bricks are subjected when manufactured. (EPA)

burner *Conical b.*, a hollow, cone-shape combustion chamber that has an exhaust vent at its point and a door at its base through which waste materials are charged; air is delivered to the burning solid waste inside the cone. Also called a *teepee burner*. (EPA) *Primary b.*, a burner that dries out and ignites materials in the primary combustion chamber. (EPA) *Refuse b.*, a device for reducing the volume of solid waste by burning.

burning area The horizontal projection of a grate, a hearth, or both. (EPA) *Effective b.a.*, the surface area of the available air spaces of the grate, at the point of contact with the material to be burned.

burning rate The mass of solid waste burned, or the quantity of heat released, divided by area and by time during incineration.

burning, suspension The burning of solid particles in a gas stream.

burnout stage The final stage of the combustion of solid waste on an incinerator grate, where the combustion of the slower-burning elements in the waste continues until a virtually organic-free clinker with a very low carbon content is left for disposal in the clinker handling system.

C

cable pullout unloading A procedure in which a landfill tractor empties a transfer trailer by pulling a cable network from the front to the rear of the vehicle.

calorific value or, especially in the USA, **heating value** The quantity of heat produced when unit mass of a material (usually a fuel) undergoes complete combustion under certain specified conditions. It is expressed in terms of kilojoules per kilogram (kJ/kg) for solid or liquid fuels, and kilojoules per cubic metre (kj/m^3) for gases. The *gross calorific value* includes the enthalpy of vaporization or, in this case, since mass is involved, the specific enthalpy of vaporization (formerly called "latent heat of vaporization"); the *net calorific value* omits it. The term "heating value" is used especially when the burning of waste is used as a means of energy production; if it is so used, *high heating* (or *heat*) *value* is used instead of gross calorific value and *low heating* (or *heat*) *value* instead of net calorific value.

calorimeter A device for measuring a change in heat content. *Oxygen-bomb c.*, a calorimeter in which a known mass of a material is ignited electrically and burned in oxygen under pressure, and the quantity of heat evolved determined (e.g., by measurement of the rise in temperature of water surrounding the bomb). Bombs vary widely in construction and method of use (some, for example, are used isothermally).

capacity, installed The handling or processing capacity of equipment or machinery installed in a plant.

capping The covering of a completed solid waste tip by waterproof material to prevent percolation through it.

car barbecue plant A high-temperature furnace specially designed to process scrap cars and produce high-grade scrap metal.

carbonaceous matter Pure carbon or carbon compounds present in the fuel or residue of a combustion process. (EPA)

carbon, activated A form of carbon that is characterized by its high adsorptive properties. It often takes the form of charcoal produced from wood or coconut shells, but it is also produced from bones and other sources. The carbon is activated by heating it to a temperature of 800-900 °C in the presence of steam, which gives it a porous internal structure, greatly increasing the surface area available for adsorption. Activated carbon (also termed "active carbon") is widely used as a decolourizing and deodorizing agent, as a gas adsorbent, and for many other purposes.

carbon dioxide recorder An instrument that continuously monitors and records the volume fraction of carbon dioxide in a mixture of gases, such as flue gas (the volume fraction is usually recorded in the form of a percentage).

carbon, fixed That part of the carbon content of a material that remains after the carbonization of the material, e.g., during the proximate analysis of a dry solid waste sample.

carbon/nitrogen ratio The relative proportions of carbon and nitrogen in organic material forming the feed for composting processes. It is generally accepted that, for satisfactory results, a C : N ratio of between 25 : 1 and 50 : 1 is required.

carry-cloth A large piece of canvas or burlap used to transfer solid waste from a residential solid waste storage area to a collection vehicle. (EPA)

carry-over Water and/or solids entrained mechanically with the steam from boilers or evaporators; the process whereby such entrainment takes place.

catchment area The area from which solid waste is received at a treatment plant, transfer station or landfill.

catch pit A pit provided in a drainage system to collect grit.

cell A zone of compacted solid waste enclosed by natural soil or cover material in a *sanitary landfill* (q.v.). See also *furnace cell*

Centralsug® system A proprietary type of pneumatic transportation system for solid waste. Of Swedish design, the Centralsug system consists of an arrangement of horizontal transporter pipes connecting vertical refuse chutes in a building or group of buildings. Refuse put into the chute is periodically conveyed pneumatically to a storage silo or incinerator at the discharge end of the transporter pipes. Valves at the foot of the chutes control the input to the transporter pipes, in which a partial vacuum is maintained by means of air pumps.

centroid of solid-waste generation The centre of a geographical area in which solid waste is collected.

cesspool or **cesspit** A tank, usually underground, for the reception of sewage, the liquid being lost mainly by seepage into the ground. (WHO) Cf. *septic tank*

cesspool emptier A tank vehicle designed to empty cesspools by suction.

charge The quantity of solid waste introduced into a furnace at one time. (EPA)

charging chamber The loading enclosure of a stationary compactor that holds material to be compressed in the compaction chamber. (POLLOCK)

charging chute A passage through which solid waste is fed into the furnace of an incinerator.

charging cutoff gate A modified charging gate used in continuous-feed furnaces that do not have high temperatures near the charging hopper. A sliding steel plate at the bottom of the charging hopper closes on a machined seat at the top of the charging chute. (EPA)

charging equipment Equipment used to charge refuse into an incinerator, e.g., hoppers, charging gates, cranes, accessories. (EAWAG)

charging gate A horizontal, movable cover that closes the opening on a top-charging furnace. (EPA)

charging hopper An enlarged opening at the top of a charging chute. (EPA)

checker work A pattern of multiple openings in a refractory structure through which the products of combustion pass to accelerate the turbulent mixing of gases. (EPA)

chemical incineration works Works for the destruction by burning of wastes produced in the course of organic chemical reactions which occur during the manufacture of materials for the fabrication of plastics and fibres and works for the destruction by burning of chemical wastes containing combined chlorine, fluorine, nitrogen, phosphorus or sulphur. (UK)

chemical oxygen demand (COD) The amount of oxygen consumed in the chemical oxidation of organic matter; a measure of the oxygen-consuming capacity of inorganic and organic matter present in water or wastewater. It is expressed as the amount of oxygen consumed from a chemical oxidant in a specific test. It does not differentiate between stable and unstable organic matter and thus does not necessarily correlate with *biochemical oxygen demand* (q.v.).

chemisorption See *adsorption*

chipper A size-reduction device having sharp blades attached to a rotating shaft (mandrel) that shaves or chips off pieces of certain objects, such as tree branches or brush.

choke beam A beam located across the discharge end of a conveyor hopper to regulate the depth of material that can pass. (BSI)

chute An essentially vertical pipe passing from floor to floor in a multi-storey building and through which refuse passes into a container chamber.

chute vent A pipe connecting a refuse chute to the open air for ventilation purposes.

classification, air A process in which the lighter components of shredded solid waste, e.g., paper and plastics, are separated from denser material. The shredded waste is fed in a controlled path through a counter-flow air stream, and separation of the component materials takes place according to their relative densities and air resistances. See also *separation; separator*

cleansing, public The specialized technology of solid waste management as applied to municipal services. The term includes solid waste storage, collection, treatment and disposal; salvage recovery; street and gully cleaning; snow removal; and ancillary services. (United Kingdom usage)

clinker Hard, sintered, or fused pieces of residue formed in a fire by the agglomeration of ash, metals, glass, and ceramics. (EPA)

clinkering 1. The act of removing clinker from a furnace. 2. The fusion of ash to form clinker.

clogging Blockage of a pipe, filter medium, etc. by solid deposits. (EAWAG)

coagulation See *aggregate*

coliform organism A Gram-negative bacillus belonging to the family Enterobacteriaceae that resembles *Escherichia coli* in its morphological and most of its biochemical characteristics. The coliform organisms are usually taken to include — in addition to *E. coli* — *Aerobacter aerogenes, Klebsiella pneumoniae,* and an ill-defined group known as the paracolon bacilli.

collection The act of removing solid waste from the central storage point of a primary source. *Alley c.*, the picking up of solid waste from containers placed adjacent to an alley. *Carryout c.*, crew collection of solid waste from an on-premise storage area using a carrying container, carry-cloth, or a mechanical method. *Contract c.*, the collection of solid waste carried out in accordance with a written agreement in which the rights and duties of the contractual parties are set forth. *Curb c.*, collection of solid waste from containers placed adjacent to a thoroughfare. *Franchise c.*, collection made by a private firm that is given exclusive right to collect for a fee paid by customers in a specific territory or from specific types of customers. *Municipal c.*, the collection of solid waste by public employees and equipment under the supervision and direction of a municipal department or official. *Private c.*, the collection of solid waste by individuals or companies from residential, commercial, or industrial premises; the arrangements for the service are made directly between the owner or occupier of the premises and the collector. (All EPA) *Separate c.*, the collection of individual components of a solid-waste stream, in order to recover or to facilitate collection and disposal. *Setout/setback c.*, the removal of full and the return of empty containers between the on-premise storage point and the curb by a collection crew. (EPA)

collection method *Daily route c.m.*, a method in which each collection crew is assigned a weekly route that is divided into daily routes. *Definite working day c.m.*, a variation of the large-route method in which definite routes are laid out and a crew assigned to each. Collection proceeds along a route for the length of time adopted for a working day. The next day, collection begins where the crew stopped the day before. This procedure continues until the whole route is covered, whereupon the crew returns to the beginning of the route. *Group task c.m.*, a method in which the responsibility for collecting on assigned routes is shared by more than one crew. Any crew that finishes a particular route works on another until all are completed. *Inter-route relief c.m.*, a method in which regular crews help collect on other routes when they finish their own. *Large route c.m.*, a method in which each crew is assigned a weekly route. The crew works each day without a fixed stopping point or work time, but it completes the route within the working week. *Reservoir route c.m.*, a method in which several crews are used to pick up on a centrally located route after having collected on peripheral routes. *Single load c.m.*, a variation of the daily route method in which areas or routes are laid out that normally provide a full load of solid waste. Each crew usually has at least two such routes for a day's work. The crew quits for the day when the assigned number of routes is completed. *Swing crew c.m.*, a method in which one or more reserve work crews go anywhere help is needed. *Variable-size crew c.m.*, a method in which a variable number of collectors is provided for individual crews, depending on the amount and conditions of work on particular routes. (All EPA)

combustion A chemical reaction in which a material combines with oxygen with the evolution of heat: "burning". The combustion of fuels containing carbon and hydrogen is said to be complete when these two elements are all oxidized to carbon dioxide and water. Incomplete combustion may lead to (*a*) appreciable amounts of carbon remaining in the ash; (*b*) emission of some of the carbon as carbon monoxide; and (*c*) reaction of the fuel molecules to give a range of products of greater complexity than that of the fuel molecules themselves (if these products escape combustion they are emitted as smoke). *Spray c.*, a process in which combustible liquids are injected into the combustion chamber of an incinerator through spray nozzles.

combustion air Air supplied to the fuel (solid waste) in a furnace to enable combustion to take place. Combustion air in a furnace consists of (*a*) *primary* or *underfire air* supplied to the underside of a firebed, and (*b*) *secondary* or *overgrate air* supplied over the firebed. For complete combustion, close regulation of the primary and secondary air is essential.

combustion chamber *Primary c.c.*, the chamber in an incinerator where waste is ignited and burned. *Secondary c.c.*, the chamber of an incinerator where combustible solids, vapours, and gases from the primary chamber are burned and fly ash is settled.

combustion, energy of The changes in heat content (i.e., the quantity of heat produced) when a substance is completely burned in oxygen, the volume being maintained constant. The heat produced by the combustion of 1 mole of the substance is the *molar energy of combustion;* that produced by the combustion of unit mass is the *specific energy of combustion*. The unit for molar energy of combustion is joule per mole (J/mol) or kilojoule per mole (kJ/mol): for specific energy of combustion it is joule per kilogram (J/kg) or joule per gram (J/g). This quantity is often called "heat of combustion".

combustion products The gases, vapours and solids that result from the combustion of a fuel.

combustion residue The solid residue (ash and clinker) resulting from combustion of material in an incinerator.

combustion system, catalytic A system in which a catalyst is introduced into an exhaust gas stream to promote the burning or oxidation of vapourized hydrocarbons or odorous contaminants; the catalyst itself remains unchanged.

comminution The reduction in size of solid waste by grinding or milling.

commissioning test One of the tests carried out under working conditions on a newly constructed plant to determine whether performance is in accordance with specification. See also *acceptance test*

compaction ratio A term often used loosely to define the performance of solid waste compaction machines. It implies that the final density of compacted waste is given by the product of input density and compaction ratio, but this is not necessarily so, as spring-back of the material takes place on release of pressure. Compaction is better described in terms of input and output volume: see *volume reduction ratio*.

compaction, surface A process whereby the dry density of surface soil is increased by the application of a dynamic load.

compaction system *Container c.s.*, an on-site system employing a power-operated compaction unit to compress refuse into containers. *Sack c.s.*, an on-site system in which power-operated rams compress the refuse into disposable sacks.

compactor *Carousel c.*, a turntable carrying refuse storage sacks, located underneath a refuse chute. A compaction device automatically compacts the material falling into the sacks, and the turntable automatically places a fresh sack under the chute when the preceding sack has been filled. *Landfill c.*, or *wheeled c.*, a mobile machine running on steel wheels fitted with special wedges or bars, used for compacting solid waste on a sanitary landfill. *Stationary c.*, a machine that reduces the volume of solid waste by forcing it into a container. (EPA)

compactor lock [One of the] mechanical connectors attached to a stationary compactor that secure the container to the compactor frame. (POLLOCK)

compactor system, stationary An integrated system designed to compact, store, transport and discharge solid waste. (POLLOCK)

compost Relatively stable decomposed organic material. (EPA) *Mature c.*, organic material that has undergone the final stage (maturing) of a composting process. This stage requires a period of several months, during which the material is aerated either by turning it over several times or by forced aeration. At the end of the maturing process, stabilization is complete. *Raw c.*, material undergoing composting. *Rough c.*, composted material that has not been screened or subjected to any other process to remove material that would be undesirable in the compost, such as glass or stones.

composting The biological breakdown of organic solids so as to stabilize them, producing a humic substance (*compost*) valuable as a fertilizer base or as a soil conditioner. *Area c.*, a composting process in which ground organic material is piled to a uniform depth on defined areas. Controlled amounts of air are distributed through ducts under each area and pass up through porous soil or a specially designed porous floor. *Mechanical c.*, a method in which the compost is continuously and mechanically mixed and aerated. (EPA) *Ventilated cell c.*, a composting method in which the compost is mixed and aerated by being dropped through a vertical series of ventilated cells. (EPA) *Windrow c.*, an open-air method in which compostable material is placed in windrows, piles, or ventilated bins or pits and is occasionally turned or mixed. The process may be anaerobic or aerobic. (EPA)

compound An enclosed area of land used for storage, e.g., a fenced area for the storage of containers for receiving solid waste.

conditioning Treatment of sludge to improve its drainage and filtration characteristics, e.g., addition of chemicals or heat treatment. (EAWAG)

conditioning tower A tower in which incinerator combustion gases are cooled before being admitted to a gas cleaning device. In the tower, the hot gases come into contact with atomized water, which is evaporated, thereby lowering the temperature of the gases. Conditioning towers may be designed for either up-flow or down-flow operation.

container A receptacle for the storage of solid waste; it may either be emptied mechanically into a collection vehicle or removed on a container-exchange basis for subsequent disposal of the contents. A container has a greater capacity than a dustbin and is commonly used at premises that generate large quantities of solid waste. *Bulk c.*, a movable container with a capacity in the range $1-9.2\,m^3$. *Communal c.*, a refuse storage container serving groups of houses or dwellings. *Individual refuse c.*, a small container in which refuse is stored while awaiting collection; in the United Kingdom it is defined as having a capacity not exceeding $0.11\,m^3$. *Refuse storage c.*, a movable refuse container with a capacity not exceeding $1\,m^3$ ($1¼\,yd^3$), in which refuse is stored while awaiting collection. (BSI)

In the USA, it may take several forms. *Carrying c.*, a receptacle of 35 to 50 gallons [132.5–189 litres] capacity, usually constructed of plastic or aluminium, that is carried by a collector in a backyard carryout service; frequently called a *tote barrel*. *Disposable c.*, plastic or paper sacks designed for storing solid waste. *Lift and carry c.*, a large container that can be lifted onto a service vehicle and transported to a disposal site for emptying; also called a *detachable c.* or *drop-off box*. *Roll-on/roll-off c.*, a large container

(20 to 40 cubic yards [15–30 m^3]) that can be pulled onto a service vehicle mechanically and carried to a disposal site for emptying. (All EPA)

container chassis A vehicle designed to carry containers.

container shelter A structure separate from the main building, at residential or commercial premises, used to house refuse storage containers.

container system, roll-on A self-contained system of loading and unloading containers on and off a vehicle chassis.

container train Small trailers hitched in series that are pulled by a motor vehicle; they are utilized to collect and transport solid waste. (EPA)

contraries Material that has to be extracted from wastepaper for recycling. *Pernicious c.,* substances that cannot readily be detected and that would interfere with the manufacture or quality of the board product, e.g., waxed laminates, vegetable parchment and non-water-soluble adhesives. *Nonpernicious c.,* e.g., string, rag, glass, metal and wood.

conveyor *Apron c.* or *plate c.,* one or more continuous chains that are supported and moved by a system of sprockets and rollers; they carry overlapping or interlocking plates that move bulk materials on their upper surface. (EPA) *Dragbar c.* or *chain c.,* a conveyor consisting of chains with interconnecting bars that move in a metal trough. The bars drag the conveyed material along the trough. *Flight c.,* a drag conveyor that has rollers interspersed in its pull chains to reduce friction. (EPA) *Residue c.,* a conveyor, usually a drag- or flight-type, used to remove incinerator residue from a *quench trough* (q.v.) to a discharge point. (EPA) *Screw c.,* a conveyor in the form of a helix that rotates in a tube. *Shuttle c.,* a conveyor with a shuttle device that can be moved along it to discharge material at any required position.

cooling sprays Water sprays directed into flue gases to cool them and, in most cases, to remove some fly ash. (EPA)

cost per tonne-minute A factor often used in cost comparisons between transfer and direct-haul operations.

cover material Coarse to fine grained material. It is commonly natural soil, that is used to cover compacted solid waste in a sanitary landfill and is free of large objects that would hinder compaction and free of organic material that would be conducive to vector harbourage, feeding and/or breeding. (EAWAG)

cover, primary Soil or other suitable compactable material spread on a layer of solid waste in order to prevent odour, insect and rodent infestation, and windblown litter. The primary cover also helps to limit the spread of fire should this occur and to reduce water percolation through the tip.

crawler tractor A tractor fitted with "crawler tracks" (i.e., endless chains of linked plates instead of wheels) through which its weight is transmitted to the ground, and used extensively for spreading and compacting solid waste in landfills.

crane *Bridge c.,* a lifting unit that can manoeuvre horizontally in two directions. *Electromagnetic c.,* a crane fitted with an electromagnetic lifting attachment that is energized and de-energized for handling ferrous material, such as scrap metal. *Grab c.,* a crane fitted with a grab; commonly used to move solid waste from storage bunkers to incinerator feed chutes or other treatment processes. *Monorail c.,* a lifting unit, suspended from a single rail, capable of moving in one horizontal direction. *Portal (jib) c.,* or *gantry c.,* a jib crane carried on a four-legged portal. The portal is built to run on rails set parallel to the quayside in the floor of a quay. Wagons and lorries can pass under the portal or the portal can pass over them. (SCOTT)

crusher, impact A machine for breaking down large or bulky items of solid waste. A heavy rotor to which projections are rigidly attached throws the material against impact plates and crushes it.

cullet Crushed glass used in glassmaking to speed up the melting of silica sand.

curing bin A container in which compost from a mechanical digestion plant is stored for 48 hours while being aerated by forced air.

cut 1. Portion of a land surface or an area from which earth or rock has been or will be excavated. 2. The distance between an original ground surface and an excavated surface. (Both EPA)

cut and cover or **cut and fill** An infrequently and incorrectly used term referring to the trench method of sanitary landfilling. (EPA)

cut-off plate A movable steel plate at the discharge end of a refuse chute, used to seal off, when required, the chute from the container or compaction unit underneath the chute.

cut-off trench A trench filled with impermeable material to prevent the passage of gas or percolate at a controlled tip site.

cyclone See *dust separator*

D

damper A hinged or sliding plate in a duct admitting air to a furnace, or within a combustion gas flue to control furnace depression. *Barometric d.*, a balanced pivoted plate set in a flue between a furnace and its stack, and actuated by the draft in the stack. If the draft becomes excessive as a result of meteorological conditions at the mouth of the stack, the damper prevents an undue rise in gas flow through the furnace. *Butterfly d.*, a plate or blade capable of rotating on an axis and installed in a duct, breeching, flue connexion or stack to regulate the flow of gases. *Guillotine d.*, an adjustable plate, utilized to regulate the flow of gases, installed vertically in a breeching. (EPA) *Sliding d.*, a plate normally installed at right angles to the flow of gas in a breeching and arranged to slide across it so as to regulate the flow.

Dano® biostabilizer system An aerobic, thermophilic composting process in which optimum conditions of moisture, air, and temperature are maintained in a single, slowly revolving cylinder that retains the compostable solid waste for one to five days. The material is later windrowed. (EPA)

deaeration The removal of air from water, usually for the control of corrosion.

decomposition, aerobic Decomposition of organic waste by the action of aerobic bacteria.

decomposition, anaerobic Decomposition of organic waste by the action of anaerobic bacteria.

deduster, wet Equipment for the removal of dust from air by passing the latter across water and through a spray zone; the dust is settled out as a sludge.

dedusting The removal of particles of dust from dust-laden air, e.g., at a refuse storage bunker.

degradation A particular type of gradual decomposition that usually proceeds in well defined stages to give products with fewer carbon atoms than the original compound. The term is often applied to decomposition resulting from the action of microorganisms.

de-inking Removal of printing ink from waste paper.

density, bulk A term frequently applied to the "apparent" mass density of a material or object, i.e., the ratio of its mass to its volume including all interstices or pores within it.

density, mass The ratio of the mass of a body to its volume. The density of a landfill is the ratio of the combined mass of the fill material and soil cover to their combined volume.

destructor A term sometimes used (incorrectly) for a solid-waste *incinerator* (q.v.).

detritus The heavier solid matter in sewage, usually mainly inorganic.

dewatering The extraction of a portion of the water present in a sludge or slurry. It implies a change from a liquid to a spadable condition, rather than concentration on the one hand or drying on the other.

dichromated oxygen consumed See *chemical oxygen demand*

digester A tank in which sludge is placed to permit digestion to occur. (EAWAG) *Fairfield-Hardy D.,*® a patented product of Fairfield Engineering Company, Marion, Ohio, which decomposes garbage, sewage sludge, industrial, and other organic wastes by a controlled continuous aerobic-thermophilic process. (EPA) *Vertical d.,* a refuse composting system or process in which the refuse is transferred each day from a given cell to one lower down in a multistorey building; the refuse passes through six such cells.

digestion The breakdown of organic substances by anaerobic action of microorganisms. *Mesophilic d.,* digestion carried out at about 20 °C. *Sludge d.,* the process whereby organic or volatile matter in sewage sludge is converted partly into gas and partly into more stable organic matter by the action of bacteria. *Wet d.,* a solid waste stabilization process in which mixed solid organic wastes are placed in an open digestion pond to decompose anaerobically. (EPA)

dilution air Air introduced over the firebed in a solid waste furnace for purposes of temperature control. In some instances, dilution air and secondary combustion air may be combined. See also *combustion air*

direct haul The transport of solid waste from the collection area to a disposal facility in collection vehicles.

discharge, horizontal A method of discharging the load from a vehicle so that the floor of the vehicle body remains in a horizontal position. On uneven ground, this helps to keep the vehicle stable.

dispersion The dilution or removal of a substance by diffusion, turbulence, etc. Technically, a two-phase system involving two substances, the first of which is uniformly distributed in a finely divided state through the second (the dispersion medium). (EPA)

disposal facility A site or plant where solid waste may be deposited for treatment or final disposal.

disposal, on-site The elimination or reduction of the volume or weight of solid waste on the property of the generator.

disposal, pit Disposal of wastes into a hole or cavity in the ground. (EAWAG)

distillation, destructive The heating of organic matter in the absence of air, with consequent decomposition of the matter and distillation of the volatile products. See also *Lantz process; pyrolysis*

dolphin A timber or steel structure at the entrance to a dock that acts as a guide to barges and ships entering the dock.

downpass or **downtake** A chamber or gas passage placed between two combustion chambers to carry the products of combustion downward. (EPA)

downtime See *outage* (2)

draft See *draught*

dragline A revolving shovel that carries a bucket attached only by cables and digs by pulling the bucket toward itself. (EPA)

dragline scraper A cable-controlled bucket for withdrawing loose material from a pile or storage bunker.

drag plate A plate beneath a travelling or chain-grate stoker used to support the returning grates. (EPA)

drain *Agricultural d.* or *field d.,* unsocketed, unglazed, earthenware, or porous concrete pipes about 3 in [7.6 cm] internal diameter, laid end to end without closing the joints so as to drain the subsoil. *French d.,* agricultural drain with the pipe surrounded by filter material like gravel, preferably by a graded filter. (Both SCOTT)

drainage 1. In general, the removal of surface water from a given area either by gravity or by pumping. Commonly applied to surface water and groundwater. 2. The area from which water occurring at a given point or location on a stream originates. In such cases, synonymous with "drainage area" and "watershed". 3. In a general sense, the flow of all liquids under the force of gravity. (All EAWAG)

draught or **draft** The difference between the pressure in an incinerator, or any component part, and that in the atmosphere; it causes air or the products of combustion to flow from the incinerator to the atmosphere. *Forced d.*, the positive pressure, created by the action of a fan or blower, that supplies the primary or secondary combustion air in an incinerator. *Induced d.*, the negative pressure created by the action of a fan, blower or ejector located between an incinerator and a stack. *Natural d.*, the draught in a furnace created by the buoyancy of the hot gases in the chimney resulting from the difference in temperature, and thus in density, between the flue gases and the atmosphere.

draught controller An automatic device that maintains a uniform furnace draught by regulating a damper.

dredging The removal of silt from the bed of a dock or river by bucket loader, grab or suction.

drop arch A structure that supports a vertical refractory furnace wall and serves to deflect gases downwards.

drop-off box See *container*

drum mill or **drum pulverizer** A long, inclined steel drum that rotates and grinds solid wastes in its rough interior; smaller ground material falls through holes near the end of the drum and larger material drops out of the end. The drum mill is used in some composting operations. (EPA)

Dulong's formula A formula for calculating the approximate heating value of a solid fuel on the basis of its ultimate analysis.

dump A land site where solid waste is disposed of in a manner that does not protect the environment. (EPA)

dumper A small, rubber-tyred, four-wheeled vehicle with a dumping hopper in front of the driver.

dumping Disposing of wastes in an uncontrolled manner. (EAWAG) *Promiscuous d.*, see *fly tipping*

dump plate A hinged plate in an incinerator that supports residue and from which residue may be discharged by rotating the plate. (EPA)

dunnage Wood, cardboard or paper used to secure raw materials or equipment in rail cars, trucks or ships.

Dusseldorf test A standard procedure for analysing the residue from a solid waste incinerator in order to determine the degree of "burn-out" of the waste, with particular reference to the putrescible content.

dust A general term denoting solid particles of different dimensions and origins, which may generally remain in suspension in a gas for some time. (Provisional ISO, 2) National standards may be more specific and include particle diameters or a definition in terms of a sieve of specified aperture size. Dust occurs in the atmosphere both naturally and as a result of the activities of man.

dustbin See *bin*

dust burden or **dust loading** The concentration of solid particles in combustion gases entering the gas-cleaning plant of an incinerator; often measured on a weight/volume basis but preferably on a weight/weight basis, which allows a more specific evaluation of emissions.

dust separator or **dust collector** An apparatus for separating solid particles from a gas stream in which they are suspended. (Provisional ISO, 2) Dust separators are widely used to remove dust and grit from stack gases and other industrial gases. They are of many different types, which may be broadly classified on the basis of whether they use or do not use a liquid to remove particles. Dust separators of the former type are known as washers.
    Non-washer types. The *cyclone,* in gas cleaning, is a dust separator or droplet separator using essentially the centrifugal force derived from the motion of the gas. (Provisional ISO, 2) A dust separator consisting of several cyclones in parallel is known as a *multicyclone*. In the *inertial separator* particles are removed by a combination of inertial and centrifugal forces resulting from a sudden change in direction of flow of the gas, together with impaction of the particles on a target. In the *electrostatic precipitator* the gas is passed between sets of electrodes across which a very high constant potential is maintained. The dust particles become charged and adhere lightly to one set of electrodes; they are removed by sharp tapping, causing them to fall into containers. Although expensive to install, electrostatic precipitators are relatively cheap to operate and are one of the most efficient means of removing fine dust from gases, although strict maintenance

is necessary to ensure unimpaired efficiency. They are widely used in power stations burning pulverized fuel, cement works, steel works, etc.

Washer types. In the *bubble washer,* the gas is bubbled through the liquid; in a *spray washer* it is passed through a chamber in which a spray of the liquid removes the dust. In the *venturi scrubber* the gas is drawn at high velocity (60-90 m/s) through a conical restriction, at the throat of which water is injected at the rate of about 1 m^3 for each 1000 m^3 of gas. The water is broken down into very fine droplets and the dust particles are thoroughly wetted; dust and water particles are subsequently removed in a cyclone. Such equipment will remove some 99% of particles smaller than 0.5 μm in diameter. Capital costs are comparatively low, but since the pressure drop across the equipment is large the power requirements are high. See also *deduster, wet*

dust tray A compartment fitted underneath the loading hopper of a refuse collection vehicle to retain dust that escapes past the compaction mechanism.

dynamic flow system The pattern of groundwater flow from a recharge to a discharge area.

E

earth blade A heavy, broad plate connected to the front of a tractor and used to push and spread soil or other material.

earth-moving machine A generic term applied to a range of mobile machines designed for excavating, moving or spreading earth and similar material, e.g., shovels, bulldozers, and scrapers. Machines of this type are used in solid waste landfill operations.

economizer A heat exchanger in a boiler system interposed between the boiler and the feed-water pump. Residual heat from combustion gases that have passed through the boiler is extracted in the economizer to raise the temperature of the boiler feed-water and thereby improve the thermal efficiency of the plant.

efficiency With regard to filters, dust separators, and droplet separators, the ratio of the quantity of particles retained by a separator to the quantity entering it (it is generally expressed as a percentage). (Provisional ISO, 2)

effluent Any fluid discharged from a given source into the external environment. (Provisional ISO, 2)

effluent seepage Diffuse discharge on to the ground of liquids that have percolated through solid waste or another medium; they contain dissolved or suspended materials. (EPA)

efflux velocity The linear velocity with which gas leaves a stack, which is equal to the volume of gas issuing from the stack mouth per second divided by the cross-sectional area of the mouth. If the efflux velocity is low, downwash is liable to occur. If several furnaces are connected to a single stack the efflux velocity is often undesirably low when some of the furnaces are not in operation, and it is an advantage to provide each furnace with a separate flue extending to the top of the stack inside the outer casing.

electrostatic precipitator See *dust separator*

elutriation A method of separating particles using the difference in apparent weight which may exist between the particles when they are suspended in a fluid. (Provisional ISO, 2) In practice, the particles are usually allowed to settle against an upward-moving flow of fluid (e.g., water or air); heavier material settles to the bottom, while fine material remains suspended and is removed with the fluid.

emission A measure of the extent to which a given source discharges a pollutant, commonly expressed either as a rate (amount per unit time) or as the amount of pollutant per unit volume of gas emitted.

emission standard A rule or measurement established to regulate or control the amount of a given pollutant that may be discharged into the outdoor atmosphere from its source. (EPA)

endothermic Of a chemical reaction, one that takes place with the evolution of heat.

equipotential lines (groundwater) Lines connecting points of equal potential.

erosion 1. The natural removal of the surface of land by the action of wind or water. 2. In a furnace or boiler plant, the wearing away of refractory or metallic surfaces by the action of moving liquids such as molten slag, or of moving gases containing abrasive dust.

evaporative tower See *conditioning tower*

evapotranspiration Amount of water transferred from the soil to the atmosphere by evaporation and plant transpiration. (WMO/UNESCO)

evase stack An expanding connexion on the outlet of a fan or in an air flow passage; its purpose is to convert kinetic energy into static pressure. (EPA)

excreta Liquid and solid waste products, human or animal. (EAWAG)

exothermic Of a chemical reaction, one that takes place with the evolution of heat.

expansion chamber See *baffle chamber*

expansion joint A joint left open in a refractory structure, or a small gap built into such a structure, so that it can expand thermally or permanently. Such joints permit sections of masonry to expand and contract freely and prevent the distortion or buckling of furnace structures under excessive expansion stresses. They are built in such a way as to permit movement of the masonry and to prevent leakage of air or gas through it.

expansion, permanent or **secondary** The permanent increase in size of some refractories at temperatures within their useful range.

extraction See *recovery*

F

fabric filter See *filter bag*

face, tip The sloping area of a landfill layer where solid waste is compacted and spread before being covered; the working face.

fallout Solid particulate material deposited from the air under gravity onto the surface of the earth.

fallout zone The surface area of the ground over which fallout occurs.

fan, forced-draught A fan used to force air into a furnace or combustion chamber, overcoming resistance, usually that of the firebed, in so doing.

fan, induced-draught A fan used to draw air from one part of a plant to another, e.g., from below the grate, through the combustion chamber to the chimney in an incineration plant.

feed chute See *charging chute*

feed hopper See *charging hopper*

feed pump The pump that provides a steam boiler with feed-water.

fermentation 1. A change brought about by a ferment, as yeast enzymes. (EAWAG) 2. Changes in organic matter or organic wastes brought about by microorganisms. (EAWAG) *Closed f.,* a process in which a part of the fermentation takes place in a closed system (digester, drum). See also *windrow composting* under *composting*

fermentation drum A long horizontal drum in which organic solid waste is subjected to a fermentation process by controlled aeration to produce material for compost. The waste is retained in the slowly rotating drum for periods of up to 6 days, and is physically broken down to a small size by attrition in the drum.

fermentation tower A vertical silo fitted with a number of decks in which organic solid waste, which has previously been ground to a small size, is subjected to aeration. Fermentation is assisted by agitation of the material during transfer from a higher deck to a lower one. The retention period is 3-5 days.

field capacity (of solid waste) Amount of water held in a soil sample after the excess of gravitational water has drained away. (WMO/UNESCO)

filling material 1. Any material deposited so as to fill or partly fill a depression. 2. The volume of material to be added. 3. An embankment. 4. Material used to raise the existing ground elevation. (All EAWAG)

filter A device or apparatus for removing solid particles from a liquid, or solid or liquid particles from a gas, in which they are suspended. The liquid or gas is passed through a filter medium, usually a granular material but sometimes finely woven cloth, unglazed porcelain or specially prepared paper.

filter bag An apparatus for removing dust from dust-laden air, employing cylinders of closely woven material which permit passage of air but retain solid particles. The term "breather bag" is deprecated. (Provisional ISO, 2) The bag may be up to 10 m in length and 1 m in diameter. An *efficiency* (q.v.) of 99-99.9% or better is quoted, depending on conditions. If a layer of chemically reactive dust is deposited on the fabric, gaseous pollutants in low concentration can be removed from stack gases (e.g., fluorine compounds can be removed from the gases evolved in the electrolytic production of aluminium). An installation containing many filter bags is known as a bag house.

filter press A press operated mechanically for partially separating water from solid materials. (EAWAG)

final cover material A layer of soil applied on top of the primary cover of the final layer of solid waste in a controlled tip to prepare the site for final restoration.

firebed The layer of material being burned on a furnace grate.

firebrick See *brick, fireclay*

fireclay A sedimentary clay containing only small amounts of fluxing impurities. It is high in hydrous aluminium silicates and is, therefore, capable of withstanding high temperatures. (EPA) See also *brick, fireclay; mortar, fireclay*

firing, auxiliary The supply of auxiliary fuel in an incinerator to assist combustion of the solid waste.

fissure A crack, break or fracture in strata or rock masses caused by faulting.

flareback A burst of flame from a furnace in a direction opposed to the normal gas glow; it usually occurs when accumulated combustible gases ignite because of sudden changes in the draught conditions or temporary blockage of the flue-gas system.

float An appliance which rests on the surface of water or sewage, usually used for registering level or for operating a switch. (WHO)

float switch A float used to control a pump or other mechanical device according to the level of water (or sewage) in a basin. (WHO)

flocculation See *aggregate*

flotation In solid waste management, the raising of suspended matter to the surface of a liquid in the form of a scum. The process depends on differences in mass density and on buoyancy effects resulting from the evolution of gas (which may be produced by chemicals, electrolysis, heat or bacterial decomposition). The scum is subsequently removed by skimming.

flue Any passage designed to carry combustion gases and entrained particulates. (EPA)

flue dust Solid particles (smaller than 100 microns [μm]) carried in the products of combustion. (EPA)

flue gas See *stack effluent*

fluid-bed separator See *dry separation* under *separation*

fluidized bed An apparatus in which a finely divided solid, usually sand, is supported and rendered fluid-like ("fluidized") by a column of rapidly moving gas, usually air, admitted from below.

fluidized-bed technique A combustion process in which heat is transferred from finely divided particles, such as sand, to combustible materials in a combustion chamber. The materials are supported and fluidized by a column of moving air. (EPA) The technique is used in a highly efficient incineration process in which shredded solid waste is fed directly into the fluidized bed. The intimate mixing of waste and sand leads to high rates of heat transfer and rapid burning.

flux In the metallurgical industries, a substance added to minerals or metals to promote fusion in a furnace; in brazing and soldering, a substance used to promote easy flow of solder and prevent oxide formation. In physics, the rate at which particles, electricity, heat, fluid, etc. flow or are transferred, expressed in terms of quantity per unit area per unit time.

fly ash Ash entrained by combustion gases. (Provisional ISO, 2) In the absence of dust separators, such ash is emitted from the stack.

fly tipping The dumping of waste or other material in unauthorized places.

fomite An inanimate object that can harbour or transmit pathogenic organisms. (EPA)

fragmentation plant A plant for the shredding or fragmentation of large pieces of metal, e.g., old refrigerators or car bodies, to produce scrap steel for recycling.

frontage factor The distance a refuse collector has to walk between individual premises when collecting from them. This affects the optimum number of collectors in a crew.

front-end loader 1. A wheeled or tracked machine with a loading shovel at the front. 2. A collection vehicle with arms that engage a detachable container, move it up over the cab, empty it into the vehicle's body, and return it to the ground.

fuel, auxiliary The fuel used to supply additional heat in a solid waste incinerator to: (*a*) dry and ignite the waste material; (*b*) maintain ignition; and (*c*) effect complete combustion of combustible solids, vapours and gases.

fuel bed See *firebed*

fume The whole of the combustion gases and the particles entrained by them (smoke). By extension, also the gases charged by particles resulting from a chemical process or a metallurgical operation. (Provisional ISO, 2) Often used in the plural for visible clouds of gases, vapours or aerosols that usually have a choking or unpleasant smell.

furnace The chamber of the incinerator into which the refuse is charged for subsequent ignition and burning. *Crematory f.*, a furnace designed to burn the carcasses of small animals. *Multiple-hearth f.*, a furnace consisting of a series of circular refractory grates, designed to burn pulverized solid waste and sewage sludge in combination.

furnace arch A nearly horizontal structure extending into a furnace and serving to deflect gases.

furnace cell Each of the separate units into which, in batch-fed incinerators, the furnace is usually divided so that charging and clinkering can be carried out independently in each such unit.

furnace volume The total internal volume of combustion chambers. (EPA)

fusion A change from the solid to the liquid state. The transition usually occurs at a sharply defined temperature (the temperature at which the two phases are in equilibrium being known as the melting point), but for some substances it occurs gradually over a range of temperature (melting range).

G

garbage Solid, organic, domestic wastes, mainly consisting of food wastes.

garbage can See *bin*

garbage grinder An electrically operated grinder installed under the kitchen sink for grinding vegetable and food wastes. The ground material passes into the sewerage system.

Garchey® system A waterborne system of solid waste removal from domestic premises. A special sink unit receives solid waste and wastewater. The waste is flushed by the water down a pipe to a receiving chamber at or below ground level, each chamber serving a number of units. The waste is subsequently removed from the chamber by a tank vehicle that separates the water for discharge to the sewer. In earlier installations, the waste slurry was partially dewatered and then burned in an on-site incinerator.

gas barrier Any device or material used to divert the flow of gases produced in a controlled tip.

gas cleaning plant See *dust separator*

gaserator A chamber in an incineration plant designed to destroy animal carcasses, condemned food and other materials that might not be completely burned if put into the main furnace. The gaserator chamber is fed by hot gases direct from the furnace.

gasification The process of converting a solid or liquid fuel into a gaseous fuel. (EPA)

gas washing or **gas scrubbing** A method of removing particulate matter from incinerator stack gases. The gases are passed through a water spray or allowed to diffuse across a series of impingement plates that are continuously washed with water. The dust particles are collected in the form of a slurry.

generation rate The amount of wastes originating from a defined activity or a defined number of waste producers per time unit. (EAWAG)

grab A device attached to a crane for lifting loose material such as solid waste. *Clamshell g.*, a grab consisting of two jaws that clamp together by the action of wire ropes, or by electrohydraulic action, when the grab is lifted. *Polyp g.*, a grab constructed of a number of pivoted curved steel plates.

grader A gas- or diesel-powered, pneumatic-wheeled machine equipped with a centrally located blade that can be angled to cast to either side. (EPA)

gradient, hydraulic The slope of the free surface of the water in any conduit, or the difference in potential between two points in a water system, divided by the distance between those points along the water flow path.

grapple A clamshell-type bucket having three or more jaws. (EPA)

grate A surface used to support fuel or solid waste while it is burning and to permit air to flow to the burning material through openings. (EAWAG) *Automatic g.*, a furnace grate that automatically ensures the continuous movement of the fuel (solid waste) through the combustion chamber by mechanical means. *Chain g.*, a *stoker* (q.v.) that has a moving chain as a grate surface; the grate consists of links mounted on rods to form a continuous surface that is usually driven by a shaft with sprockets. *Continuous g.*, a type of incinerator grate that is charged intermittently at the feed end with solid waste and moves the burning material continuously towards the discharge end, from where clinker is removed as a continuous automatic process. *Dump g.*, see *dump plate*. *Incinerator g.*, a mechanically operated grate for burning solid waste in a furnace and discharging the residue. *Inertial g.*, a stoker consisting of a fixed bed of plates that is carried on rollers and activated by an electrically driven mechanism; it draws the bed slowly back against a spring and then releases it so that the entire bed moves forward until stopped abruptly by another spring. The inertia of the solid waste carries it a short distance along the stoker surface, and the cycle is then repeated. *Mechanical g.*, any kind of incinerator grate that does not require manual stoking. *Oscillating g.*, a stoker, the entire grate surface of which oscillates to move the solid waste and residue over the grate surface. *Reciprocating g.*, a grate arranged so that alternate sections undergo a slow reciprocating motion in a horizontal direction; each forward movement pushes the burning material further along the grate surface. *Rocker g.*, a grate fitted with alternate fixed and pivoted sections; the latter, when rocked on their pivots, lift and move the burning waste forwards, at the same time exposing fresh surfaces of the material to combustion. *Shredder g.*, the section of a shredder or hammermill through which shredded or pulverized material passes. The grate usually consists of a number of bars spaced according to the maximum particle size of the shredded material required. *Travelling g.*, a stoker that is essentially a moving chain belt carried on sprockets and covered with separated small metal pieces called keys. The entire top surface can act as a grate while moving through the furnace but can flex over the sprocket wheels at the end of the furnace, return under the furnace, and re-enter the furnace over sprocket wheels at the front.

gravel Aggregate consisting of rock fragments in the general size range 2–65 mm in diameter.

grid bars Cast steel bars forming a grid at the base or outlet of a hammermill pulverizer, through which the pulverized material passes when sufficiently reduced in size.

groundwater The water contained in porous underground strata as a result of infiltration from the surface.

groundwater discharge area An area of low *groundwater potential* (q.v.), where water appears at the surface of the ground or enters a collector.

groundwater lowering The artificial local lowering of the water table to enable work to be carried out in excavations in the dry. (WHO)

groundwater potential The potential energy at a point in a groundwater system; expressed as the elevation of the water table above sea level.

groundwater runoff That part of the runoff which has passed into the ground, becomes groundwater, and is discharged into a stream channel as spring or percolation water. (WMO/UNESCO)

grouser A ridge or cleat that extends across a crawler tractor track to improve its traction. (EPA)

grout A cementing or sealing mixture of cement and water to which sand, sawdust, or other fillers may be added. (EPA)

H

hammermill A broad category of high-speed equipment that uses pivoted or fixed hammers or cutters to crush, grind, chip or shred solid wastes. (EPA) *Vertical-swing h.*, a hammermill fitted with hammers pivoted on a horizontal shaft.

handling The functions associated with the movement of solid waste materials after creation, excluding storage, processing and ultimate disposal.

hardness (of water) The presence of dissolved calcium and magnesium salts in water. Such salts react with sodium soaps to form insoluble soaps that have no detergent properties. Hardness can lead to scale formation in hot-water or boiler systems.

hardpan A hardened, compacted, or cemented soil layer. (EPA)

haul distance 1. The distance a collection vehicle travels from its last pickup stop to the solid waste transfer station, processing facility, or sanitary landfill. 2. The distance a vehicle travels from a solid waste transfer station or processing facility to a point of final disposal. 3. The distance that cover material must be transported from an excavation or stockpile to the working face of a sanitary landfill. (All EPA)

haul time The elapsed or cumulative time spent transporting solid waste between two specific locations. (EPA)

hearth *Drying h.*, a surface in an incinerator on which material with a moisture content is partially dried before being burnt. Hot combustion gases are passed over the material to dry it. *Solid h.*, a solid surface of refractory material without air gaps, used to support material during drying, ignition or combustion. Solid hearths are employed where the material to be burned consists of sludges, viscous liquids or solids that melt below ignition temperature.

heat, available The quantity of useful heat produced per unit of fuel if it is completely burned; the heat values of the dry flue-gas and water vapour are deducted. (EPA)

heat balance The balance between all forms of heat input and output of an incinerator, usually on an hourly basis.

heat, conduction of The transfer of heat through a body in which a temperature gradient exists, as a result of molecular interaction. The mechanism of heat transfer varies, depending on the nature of the body (liquid, gas, electrically conducting solid or dielectric solid).

heat exchanger A device that transfers heat from one flue to another without allowing them to mix. (EPA)

heating, central station or **district** The heating of all buildings and dwellings within a given area from a single installation, usually by means of superheated steam. The installation may be set up specially for the purpose, or it may use the waste heat from an existing installation such as a power station or municipal incinerator. Since pollution from a single large plant can be controlled much more easily than that from a multitude of small sources, central station heating can be a valuable aid in achieving clean air.

heat of combustion See *combustion, energy of*

heat release rate The quantity of heat produced per unit furnace volume per unit time during complete combustion of a fuel.

heat value, heating value See *calorific value*

hectare A unit of area, equal to $10\,000\,m^2$. The hectare is not an SI unit, but it has been accepted by the Conférence général des Poids et Mesures for use with the SI for a limited time. Symbol: ha

herbicide A substance or mixture of substances that destroys plant life. In common usage, the term is often restricted to herbicides that show selective action in killing noxious plants (weed-killers).

high heating (or **heat**) **value** See *calorific value*

hog cholera A virus disease associated with the feeding of raw garbage to swine.

hog feeding The utilization of heat-treated food wastes as a livestock feed. (EPA)

hopper A device into which refuse is placed and from which it is projected into a chute or sometimes directly into a refuse container. (BSI) *Bifurcated h.*, a chamber divided into two chutes by a hinged plate that can be moved so as to direct material to either side. *Reception h.,* a pit or bunker in which solid waste is received and stored pending subsequent handling or treatment.

hot spot A localized area of high temperature in a fuel bed in a furnace.

humus 1. Organic matter present in the soil and decomposed to such an extent that its original structure cannot be determined; generally colloidal, and dark brown in colour. 2. Flocculent settleable material contained in the effluent of a percolating filter.

hydrogenation The chemical combination of hydrogen with another substance, usually by the action of heat and pressure in the presence of a catalyst. It is widely used in the petroleum industry.

I

ignition arch An arch or surface in a refractory furnace located over a fuel bed so as to radiate heat and accelerate combustion.

ignition point 1. The temperature to which a substance must be heated before combustion can take place. 2. The lowest temperature at which self-sustained combustion of a given substance takes place (in the absence of an external source of ignition such as a flame).

Immission A German term for which there is no simple English equivalent. In the Federal Republic of Germany, *Immissionen* are legally defined as "air pollutants, noise, vibrations, light, heat, radiation and analogous environmental factors affecting human beings, animals, plants or other objects". They are to be distinguished from emissions (*Emissionen*), which are defined as "air pollutants, noise, vibrations, light, heat, radiation and analogous phenomena originating from an installation". (FRG)

impermeability The state of being impermeable; the absence of voids capable of allowing the passage of fluids. See also *permeability*

impervious See *permeability*

incineration The controlled burning of solid, liquid or gaseous combustible wastes so as to produce gases and residues containing little or no combustible material. *Central i.,* incineration at a plant that serves a community or part of a community. *Fluidized-bed i.,* the burning of shredded or pulverized solid waste by the *fluidized-bed technique* (q.v.). *On-site i.,* incineration of solid wastes at the place where they are generated. *Refuse i.,* the process of reducing combustible waste to an inert residue by high-temperature burning.

incinerator An apparatus in which wastes are burned and in which all the factors affecting combustion, namely temperature, retention time, turbulence and combustion air, can be controlled. *Cell-type i.,* an incinerator whose grate areas are divided into cells, each of which has its own ash drop, underfire air control, and ash grate. (EPA) *Central i.,* a conveniently located facility that burns solid waste collected from many different sources. (EPA)

Chute-fed i., an incinerator that is charged through a chute that extends two or more floors above it. (EPA) *Continuous-feed i.*, an incinerator into which solid waste is charged almost continuously to maintain a steady rate of burning. (EPA) *Controlled-air i.*, an incinerator with two or more combustion areas in which the amounts and distribution of air are controlled. Partial combustion takes place in the first zone and gases are burned in a subsequent zone or zones. (EPA) *Direct-fed i.*, an incinerator that accepts solid waste directly into its combustion chamber. (EPA) *Domestic i.*, a small incinerator, generally gas-fired, designed to burn domestic-type solid waste from individual houses or other small buildings. *Flue-fed i.*, an incinerator that is charged through a shaft that functions as a chute for charging waste and has a flue to carry the products of combustion. (EPA) *Hospital i.*, an incinerator designed to burn hospital wastes, e.g., surgical dressings and biological material. *Industrial i.*, an incinerator designed to burn a particular industrial waste. (EPA) *Multiple-chamber i.*, an incinerator consisting of two or more chambers, arranged as in-line or retort types, interconnected by gas passage ports or ducts. (EPA) *Municipal i.*, an incinerator designed and used primarily to burn residential and commercial solid waste. *On-site i.*, an incinerator designed to burn waste at the premises where it arises. *Open-pit i.*, an open-topped incinerator provided with a system of closely spaced nozzles that direct a stream of high-velocity air over the burning zone. *Retort-type i.*, a multiple-chamber incinerator in which the gases travel from the end of the ignition chamber, then pass through the mixing and combustion chamber. (EPA) *Slag-forming i.*, a solid-waste incinerator that operates at temperatures high enough to melt the incombustible components. The slag, after quenching, is in the form of a sterile granulated material.

incinerator capacity The quantity of solid waste that can be processed by an incinerator in a given time under certain specified conditions, usually expressed in terms of mass per 24 hours. *Design i.c.*, the capacity for which an incinerator is designed. *Firm i.c.*, the capacity of an incinerator when its largest independent unit is not operating. *Rated i.c.*, the capacity of which an incinerator is capable.

Indore process An anaerobic composting method that originated in India. Organic wastes are placed in alternate layers with human or animal excreta in a pit or pile. The piles are turned twice in six months and drainage is used to keep the compost moist. (EPA)

inertial separator See *dust separator*

infiltration The process whereby some precipitation flows through the surface of the ground. (EPA)

infiltration air Air that leaks into the chambers or ducts of an incinerator. (EPA)

inoculum Microorganisms placed in a culture medium, soil, compost, etc. (EPA)

insulation, plastic An insulation that is sufficiently plastic when mixed with water to enable it to adhere to outer furnace walls or to be placed over furnace arches.

interflow That portion of precipitation that infiltrates into the soil and moves laterally under its surface until intercepted by a stream channel or until it resurfaces down-slope from its point of infiltration. (EPA)

ion exchange The exchange of ions of the same sign of electric charge between a solution and an inorganic or organic solid or liquid called an *ion exchanger*. The ion exchanger is considered to be insoluble in the solution.

J

jack, stabilizing One of the devices used to stabilize a vehicle chassis while containers are picked up or lowered.

K

K-factor A term sometimes applied to *thermal conductivity* (q.v.).

kiln, rotary A cylinder for the combustion of solid waste, that rotates on an inclined axis, thus causing the burning material to move in a slow, cascading forward motion.

kinetic techniques Techniques of doing heavy manual work with the minimum of muscular effort, based on the principle that the arms are mainly concerned in transmitting power that originates in the effective use of body weight.

L

landfill blade A *U-blade* (q.v.) with an extension on top that increases the volume of solid waste that can be pushed and spread, and protects the operator from any debris that may be thrown out.

landscaping The process of changing the natural shape and features of land so as to make it more attractive, e.g., by adding lawns, trees, bushes, etc. In this context, usually related to sanitary landfill.

Lantz process A destructive distillation technique, in which the combustible components of solid waste are converted into combustible gases, charcoal, and a variety of distillates. (EPA) See also *distillation, destructive; pyrolysis*

leachate Liquid from a landfill containing dissolved solids either already in solution in the waste or subsequently extracted and dissolved by water passing through it.

leaching 1. The removal of soluble constituents from soils or other material by percolating water. 2. The removal of salts and alkali from soils by abundant irrigation combined with drainage. 3. The disposal of a liquid through a non-watertight artificial structure, conduit, or porous material by downward or lateral drainage, or both, into the surrounding permeable soil. (All EAWAG)

ledge plate A plate that is adjacent to or overlaps the edge of a stoker. (EPA)

lift In a sanitary landfill, a compacted layer of solid wastes and the top layer of cover material. (EPA)

lining The material used on the inside of a furnace wall; usually of high-grade refractory tiles or bricks or a plastic refractory material. (EPA) *Monolithic l.,* a refractory lining or construction made in large sections on site; conventional layers and joints of brick construction are not used. (EPA)

lipid In general, any fat or related material. The term "lipid" is preferred to "lipoid".

load-bearing resistance (refractory) The degree to which a refractory resists deformation when subjected to a specified compressive load at a specified temperature and time. (EPA)

loading line The height above ground level of that part of a refuse-collection vehicle body over which the refuse is loaded.

loading system, dustless A mechanical system of emptying bins into a collection vehicle without permitting escape of dust.

load, rated The maximum load that a crane is designed to handle safely. (EPA)

loam A soft, easily worked soil containing sand, silt, and clay. (EPA)

low heating (or **heat**) **value** See *calorific value*

lysimeter A device used to measure the quantity or rate of water movement through or from a block of soil or other material, such as solid waste, or used to collect percolated water for quality analysis. (EPA)

M

machine coding and switching A system of sensors and mechanical devices for coding and removing specific classes of material from inhomogeneous solid waste for the purpose of recycling, developed on an experimental basis at Massachusetts Institute of Technology.

manure Primarily the excreta of animals; may contain some spilled feed or bedding. (EPA)

map, topographic A map indicating surface elevations and slopes. (EPA)

materials balance The balance between the weights of materials entering and leaving a process or a system.

materials, secondary Materials other than primary or raw materials used for manufacturing processes. A term applied to recycled materials.

membrane barrier Thin layer of material impermeable to the flow of gas or water. (EPA)

metal, heavy Metallic element with high atomic number, e.g., cadmium, lead or mercury. Heavy metals can have toxic effects even in small concentrations.

mixing chamber A chamber usually placed between the primary and secondary combustion chambers of an incinerator; the products of combustion are thoroughly mixed there by turbulence that is created by increased velocities of gases, checker work, or turns in the direction of the gas flow. (EPA)

moisture content The quantity of water present in soil, wastewater sludge, industrial waste sludge, and screenings, usually expressed in percentage of wet weight. (EAWAG) The weight loss (expressed in percent) when a sample of solid waste is dried to a constant weight at a temperature of 100 °C to 105 °C. (EPA)

moisture content, percentage

1. Wet basis: $\dfrac{100 \times \text{(water content of sample)}}{\text{dry weight of sample} + \text{water content of sample}}$

2. Dry basis: $\dfrac{100 \times \text{(water content of sample)}}{\text{dry weight of sample}}$

moisture-holding capacity See *field capacity*

mortar, fireclay A mortar made of high-fusion-point fireclay and water. It is often used to fill joints in refractory walls to stop air or gas leaks without forming a strong bond. (EPA)

mortar, refractory A finely ground material that, when it dries, develops a strong bond between materials and can withstand very high temperatures. *Air-setting r.m.* or *cold-setting r.m.*, a refractory mortar that sets without the application of heat. *Heat-setting r.m.* or *hot-setting r.m.*, a mortar in which the bond is developed by the action of relatively high temperatures, which vitrify part of its constituents. (EPA) *Hydraulic-setting r.m.*, a mortar that hardens or sets as a result of hydration, a chemical reaction with water. In an incinerator, the water in the mortar evaporates and a ceramic bond develops when the working furnace temperature is applied. (EPA)

motive unit The tractor unit of an articulated vehicle.

MPL system A system used at a transfer station, in which the solid waste is compacted in a horizontal cylinder by a hydraulic ram and then transferred to a special vehicle with a cylindrical body. MPL stands for "maximum pay load", as the equipment was designed to achieve the maximum permissible load for road vehicles under United Kingdom regulations.

mudflats Areas of mud that do not support any vegetation and are at times covered by water.

mulch A layer of organic material applied to the surface of the ground to retain moisture in it or in the roots of plants.

mulching The spreading of leaves, straw, or other loose material on the ground to prevent erosion, evaporation, freezing of plant roots, etc. (EAWAG)

multicyclone See *dust separator*

N

night soil Solid and liquid human excrement collected from pails or chemical closets at premises not connected to a sewer or cesspool.

nitrification The conversion of ammonia to nitrite and nitrate by microorganisms under aerobic conditions.

nitrogen cycle A series of processes in which atmospheric nitrogen is converted into nitrates in soil and other nitrogenous compounds in plants, from which — directly, by way of animals, or through the intermediate stage of coal — ammonium compounds are formed in the soil. These ammonium compounds eventually break down to return the nitrogen to the atmosphere or to another part of the cycle.

normal temperature and pressure (NTP) An agreed set of standard conditions used in the reporting of gas volume measurements. Normal temperature and pressure refer to a temperature of 273.15 °K (0 °C) and a pressure of 101.325 kPa (1 atm). The term "standard temperature and pressure" (STP) was formerly used, but it sometimes referred to a different set of standard conditions.

O

ocean disposal The deposition of waste into an ocean or estuary.

odour threshold In principle, the lowest concentration of an odorant that can be detected by a human being. In practice, a panel of "sniffers" is normally used and the threshold taken as the concentration at which 50% of the panel can detect the odorant (although some workers have also used 100% thresholds).

opacity rating A measure of the opacity of an emission; the degree of obscuration (of the vision of an observer) that is equal to the apparent obscuration produced by smoke of a given Ringelmann number. See also *smoke chart*

orange-peel bucket See *grapple*

organic content Synonymous with *volatile solids* (q.v.), except for small traces of some inorganic materials, such as calcium carbonate, that lose weight at temperatures used in determining volatile solids. (EPA)

Orsat apparatus An apparatus for the volumetric analysis of gases, usually flue gases. The apparatus is usually used to determine carbon monoxide, carbon dioxide, and oxygen; it may be designed to determine hydrogen, nitrogen (by difference), and other gases (such as methane and ethane) also.

outage 1. The shutdown of plant for maintenance or repair. 2. The length of time the plant is out of service for such reason.

outline design stage A project design developed in sufficient detail to enable feasibility to be assessed and a cost estimate made.

overband magnet or **overband separator** An electromagnet placed transversely above a conveyor carrying solid waste or incinerated residue for the purpose of recovering ferrous metals.

overgrate air or **overfire air** Air supplied in a furnace above the firebed in order to oxidize the gases in the combustion chamber. It may also have the purpose of regulating the furnace temperature. See also *combustion air*

overhead magnet See *overband magnet*

overload capacity The handling or treatment capacity of a plant in excess of the designed capacity, and which can be used for short periods without detriment to the plant.

oxidation A loss of electrons, leading to an increase in valence number (originally the term meant combination with oxygen, but it is now given this broader meaning). *Wet o.*, a method of sludge disposal that involves the oxidation of sludge solids in water suspension and under increased pressure and temperature. (EAWAG)

oxidation pond A basin used for retention of wastewater before final disposal, in which biological oxidation of organic material is effected by natural or artificially accelerated transfer of oxygen to the water from air. (EAWAG)

P

packaging, one-way or **one-trip** Containers, bottles or other forms of packaging intended to be discarded as solid waste when empty.

pallet A support for piles or bales of material designed to facilitate mechanical handling. It may be disposable or returnable.

pasteurization A procedure for the destruction of pathogenic and fermentative bacteria in food and food products. It involves raising the temperature of the product to a specified level and maintaining it at that level for a definite period of time, followed by cooling within a specified time interval.

paper, mixed A term used in the wastepaper trade for mixed waste paper and board, screen-sorted or sorted over an endless band for the removal of *contraries* (q.v.) not suitable for conversion by the cold repulping process.

payload The weight of material carried by a loaded or partly loaded vehicle.

PCB See *polychlorinated biphenyl*

peep door A small door or hole in an incinerator, through which combustion can be observed. (EPA)

percolate See *leachate*

percolation The movement of water through a permeable stratum. (WHO)

permeability In groundwater hydrology, the property of a material that permits the passage of water through it at an appreciable rate at normal pressures. A material that permits perceptible passage of water is said to be *permeable;* a material that does not is said to be *impermeable* (the terms "pervious" and "impervious", respectively, are sometimes used). See also *impermeability*

pervious See *permeability*

pH control A process (usually automated) whereby the pH of a liquid is adjusted to, or maintained at, a certain preselected value (or within preselected limits). In solid waste management the process is used for the neutralization of acidic or alkaline sludges prior to disposal.

pH value A measurement of the presence of hydrogen ions in a solution used as an indicator of acidity (pH $<$ 7) or alkalinity (pH $>$ 7).

physisorption See *adsorption*

picking belt A conveyor on which solid waste is spread to enable materials to be removed manually, either for recycling or because they would interfere with subsequent processing of the waste.

plant, auxiliary Plant and equipment in a solid waste treatment plant, other than the principal processing equipment, e.g., in incinerators, grabs, fans and pumps.

plant commissioning The process of bringing a new plant into operation after its completion.

plant rating The manufacturer's stated performance figures for a treatment plant. Rating figures may relate to continuous, maximum or average performance.

platen A plate in a solid waste compaction machine that exerts pressure on the waste.

platform scale See *weighbridge*

plenum chamber A chamber beneath an incinerator grate from which air is drawn for distribution through the grate and the firebed.

plume The gases issuing from a stack so long as they continue to form a stream of gas and do not become completely dispersed in the surrounding air. Near the stack the plume is often visible owing to the water droplets, dust, or smoke that it contains, but it often persists downwind long after it has become invisible to the eye (it can, however, be detected and followed with suitable instruments).

pollutant Any undesirable solid, liquid, or gaseous matter in a gaseous or liquid medium. (Provisional ISO, 2) For the meaning of "undesirable" in pollution contexts, see *pollution. Primary p.,* a pollutant emitted into the atmosphere from an identifiable source. *Secondary p.,* a pollutant formed by chemical reaction in the atmosphere.

pollution The introduction of pollutants into a liquid or gaseous medium, the presence of pollutants in a liquid or gaseous medium, or any undesirable modification of the composition of a liquid or gaseous medium. (Provisional ISO, 2) For purposes of pollution control, an "undesirable modification" is one that has injurious or deleterious effects.

polychlorinated biphenyl (PCB) Any of a series of organochlorine compounds containing two linked phenyl rings and a variable proportion of chlorine. Most commerical products contain several different isomers, and some contain polychlorinated terphenyls also. They are widely used in the electrical industry, particularly in transformers, and have also been used as lubricants, hydraulic fluids, plasticizers, flame retardants, etc. They are very stable and have become widespread in the environment, and considerable concern has been expressed over their effects on health.

polyethylene A polymer of ethylene (or, in systematic nomenclature, of ethene, the polymer then being termed polyethene). The term is also applied loosely to polyethylene plastics, which are widely used for electric cable insulation, laboratory equipment, and many other purposes.

poly(vinyl chloride) A polymer of vinyl chloride, frequently referred to by the initials PVC. The term is also applied loosely to poly(vinyl chloride) plastics, which are widely used for many different purposes. (The approved way of writing the term is as in the heading to this entry; "polyvinyl chloride", though often seen, is chemically incorrect.)

porosity Ratio of the volume of the interstices in a given sample of a porous medium, e.g., soil, to the gross volume of the porous medium, inclusive of voids. (WMO/UNESCO)

Prat® process A refuse-composting process in which unsorted refuse is fermented in cells into which air is injected.

precipitation The thickness of the layer of water which accumulates on a horizontal surface, as the result of one or more falls of precipitation, in the absence of infiltration or evaporation, and if any part of the precipitation falling as snow or ice were melted. (WMO) It is usually measured in terms of depth divided by time, e.g., millimetres per day or per year.

prefeasibility study An examination of the range of possible methods of satisfying a defined operational need, using minimum resources.

preheater, air A heat exchanger through which air passes and is heated by a medium at a higher temperature, e.g., hot combustion gases. In a boiler or furnace, combustion air may be preheated in such a device.

primary air See *underfire air*

priming See *carry-over*

probe A tube with a special nozzle, used for sampling combustion gases in an incinerator.

PSI An incorrect abbreviation for pound-force per square inch. The correct abbreviation for this (non-SI) unit is $lbf\,in^{-2}$ or lbf/in^2. Conversion factor: $1\,lbf/in^2 = 6.9\,kPa$

puddling The mixing of water with clay or chalk to increase its plasticity. Puddled clay or chalk is used to line the base of landfill sites so as to prevent leachate from the solid waste penetrating strata and contaminating groundwater.

pulp A mixture of ground up, moistened cellulose material, as wood, linen, rags, etc., from which paper is made. (EAWAG)

pulverization The crushing, shredding or grinding of solid waste into small pieces, thereby making it more homogeneous. *Wet p.*, a pulverization process in which water is added to solid waste to soften the paper and cardboard.

purification The treatment of water (or sewage) for the removal of harmful or undesirable substances. (WHO)

push pit A storage system sometimes used in stationary compactor transfer systems. A hydraulically powered bulkhead that traverses the length of the pit periodically pushes the stored waste into the hopper of a compactor. (EPA)

PVC See *poly(vinyl chloride)*

pyrolysis The decomposition of organic material by heat in the absence of or with a limited supply of oxygen. Pyrolysis produces combustible gas, oil, tar and solid residue in proportions depending on the amount of air present, reaction time, temperature and pressure. See also *distillation, destructive; Lantz process*

pyrometer An instrument for measuring or recording temperatures. It may (*a*) match the intensity of radiation at a single wavelength from a tungsten filament with the intensity of the radiation at the same wavelength (optical pyrometer), or (*b*) measure the intensity of radiation at all wavelengths emitted by a material having a high temperature (radiation pyrometer).

pyrometric cone equivalent (PCE) An index to the refractoriness of a material; it is obtained by a test that provides the number of a standard pyrometric cone that is closest in its bending behaviour to that of a pyrometric cone made of the material when both are heated in accordance with the ASTM Standard Method of Test for Pyrometric Cone Equivalent of Refractory Materials. (EPA)

Q

quenching The sudden cooling of a material or the abrupt halting of a process or reaction.

quench trough A water-filled trough into which burning residue drops from an incinerator furnace. (EPA)

R

rabble arms The radial arms in a multiple-hearth furnace that act as scrapers to transport material across the hearths.

ram *Double-acting r.*, a hydraulic ram on either side of which pressure can be applied. *Feeder r.*, a hydraulic ram located at the base of a furnace charging chute, which pushes the charge (solid waste) on to the furnace grate. *Multi-stage r.*, a hydraulic ram in which the piston or inner tube consists of two or more concentric tubes; this allows greater extension of the piston than is practicable with a single-stage ram. *Single-acting r.*, a hydraulic ram in which pressure can be applied in one direction only; it has, therefore, to be returned to the closed position by some external force. *Single-stage r.*, a hydraulic ram with a piston or inner tube of fixed length.

rainfall See *precipitation*

ramp system A system of moving containers onto and off a vehicle chassis. As the vehicle backs between fixed ramps, the container is automatically lifted from chassis to ramps and vice versa.

rasp A machine for grinding solid waste. Slowly rotating arms grind the material on horizontal steel plates until the particle size is reduced sufficiently for it to pass through holes in the plates.

rasper A grinding machine in the form of a large vertical drum containing heavy hinged arms that rotate horizontally over a rasp-and-sieve floor. (EPA)

recharge area (groundwater) An area of high *groundwater potential* (q.v.), where infiltration exceeds groundwater losses.

recharging The addition, by natural or artificial means, of water to an underground aquifer. (WHO)

recirculation The return of a fully or partially treated liquid to an earlier stage of the process. (WHO)

reclamation 1. The improvement of land or its recovery from sea or swamp. (WHO) 2. The process of collecting and segregating wastes for reuse. See also *recovery; recycling*

recovery The process of obtaining materials or energy resources from solid waste. (EPA) See also *reclamation; recycling*

recycling A term embracing the recovery, return and reuse of scrap or waste material for manufacturing or resource purposes. *Direct r.*, the use of a recovered material for manufacture of a similar product. *Indirect r.*, the use of recovered material for manufacturing a different product or one of less critical specification. *Thermal r.*, conversion of the waste into energy, with or without by-products.

reduction 1. In chemistry, the addition of electrons, leading to a decrease in valence number (originally the term meant the removal of oxygen or the addition of hydrogen, but it is now given this wider meaning). 2. Conversion to a finer state, as the breaking down of solid waste into smaller pieces or particles by mechanical means.

refractory Nonmetallic substances used to line furnaces because they can endure high temperatures. In addition, they should normally be able to resist one or more of the following destructive influences: abrasion, pressure, chemical attack, and rapid changes in temperature. *Castable r.*, a hydraulic-setting refractory, suitable for casting or being pneumatically formed into heat-resistant shapes or walls. *High alumina r.*, a refractory product containing 47.5 percent more alumina than regular refractories. *Plastic r.*, a blend of ground fireclay materials in a plastic form, that is suitable for ramming into place to form monolithic linings or special shapes. It may be air-setting or heat-setting and is available in different qualities of heat resistance. (All EPA)

refuse A generic term covering all kinds of solid waste. *Market r.*, solid waste generated in food markets or at street markets. *Milled r.*, solid waste that has been mechanically reduced in size. *Municipal r.*, solid waste generated at domestic or other residential premises and at commercial premises, such as shops. *Raw r.* or *crude r.*, solid waste in its "as collected" state, before it has been subjected to any treatment process.

refuse reduction plant See *treatment plant*

refuse storage chamber A compartment containing one or more containers into which refuse is discharged. (BSI)

reinjection Reintroduction of fly ash into a furnace to burn out all the combustibles. (EPA)

reject An item of solid waste that passes through a pulverizer without being crushed or reduced in size.

rendering A process of recovering fatty substances from animal parts by heat treatment, extraction, and distillation. (EPA)

reprocessing The action of changing the condition of a secondary material. (EPA)

reseeding See *seeding*

residence time The time that elapses between the entry of a substance into a furnace and the exit of burnt-out residue from it.

residue The material remaining after combustion of wastes such as ash, slag, or cinders. Also materials extracted from a liquid or gas stream. (EAWAG) *Sewage-treatment r.,* coarse screenings, grit, or sludge from wastewater treatment units. (EPA)

resource recovery See *recycling*

resources, recoverable Materials that still have useful physical or chemical properties after serving a specific purpose and can, therefore, be reused or recycled for the same or other purposes. (EPA)

respiration, aerobic Respiration in the presence of free (i.e., gaseous or dissolved) oxygen.

riddlings The fine ash that falls between furnace grate sections and through air passages in the grate.

Ringelmann chart See *smoke chart*

riparian rights Rights of the landowner to water on or bordering his property; they include his right to prevent upstream water from being diverted or misused. (EPA)

roll crusher A pair of steel rolls rotating at different speeds that crush glass and other hard materials in pulverized solid waste.

roller track A series of cylindrical steel rollers set in parallel frames that act as a free-running conveyor for materials such as bales of metal or paper.

ropeway, aerial A line of towers carrying steel ropes that serve as tracks for buckets for transporting material, where surface transportation would be inconvenient or impracticable.

rubbish 1. In the United Kingdom, synonymous with refuse. 2. In USA usage, a general term for solid waste, excluding food waste and ashes taken from residences, commercial establishments and institutions.

rubble Broken pieces of masonry and concrete. (EPA)

runoff That part of precipitation that flows towards the stream on the ground surface (surface runoff) or within the soil (subsurface runoff or interflow). (WMO/UNESCO)

S

sack holder A metal frame to which a disposable solid waste storage sack is attached. A hinged lid is attached to the frame, and some sack holders incorporate metal guards to prevent access to the sack by animals.

sack system A system of solid waste storage in plastic or paper sacks that are removed and disposed of with the waste.

saltings Grassed areas covered by tidal water at some or all high tides.

salvage Material recovered from solid waste for the purpose of recycling.

salvaging The controlled removal of waste materials for utilization. (EPA) See also *reclamation; recovery; recycling*

sampling The process whereby a comparatively small quantity of a material, e.g., solid waste, is obtained that is representative of the whole.

sampling hole See *borehole*

sanitary landfill A site where solid waste is disposed of using sanitary landfilling techniques. (EPA)

sanitary landfilling An engineered method of disposing of solid waste on land in a manner that protects the environment, by spreading the waste in thin layers, compacting it to the smallest practical volume, and covering it with soil by the end of each working day. (EPA)

sanitary landfilling method *Area s.l.m.,* a method in which the wastes are spread and compacted on the surface of the ground and cover material is spread and compacted over them. *Quarry s.l.m.,* a variation of the area method in which the wastes are spread and compacted in a depression; cover material is generally obtained elsewhere. *Ramp s.l.m.,* another variation of the area method in which a cover material is obtained by excavating in front of the working face. A variation of this method is known as the *progressive slope sanitary landfilling method. Trench s.l.m.,* a method in which the waste is spread and compacted in a trench. The excavated soil is spread and compacted over the waste to form the basic cell structure. *Wet area s.l.m.,* a method used in a swampy area where precautions are taken to avoid water pollution before proceeding with the area landfill technique. (All EPA)

saprogenic Causing or resulting from decay.

saturation A state in which the concentration of solute in a solution is identical to that which would exist in a similar solution in equilibrium with undissolved solute at a given temperature and a given pressure.

scavenger One who participates in the uncontrolled removal of materials at any point in the solid waste stream. (EPA)

scavenging The manual sorting of solid waste at landfills to remove usable material.

scavenging process A process or mechanism — physical, chemical, or biological — that results in the removal of pollutants from the atmosphere (e.g., the removal of suspended particulate matter by rain).

scooper, hydraulic A self-propelled crawler vehicle equipped with hydraulically operated arms that lift, empty, and place containers carried on a transfer trailer bed. (EPA)

scooter See *vehicle, satellite*

scrap Discarded or rejected material or parts of material that result from manufacturing or fabricating operations and are suitable for reprocessing. (EPA) *Heavy s.,* scrap metal, such as cast iron or heavy steel section. *Light s.,* scrap metal such as sheet steel and light steel section, e.g., the casings of old refrigerators or domestic cookers, galvanized tanks, and steel drums.

scraper 1. Equipment for the removal of sludge, etc., from sedimentation basins after treatment. 2. A self-propelled or towed machine that can scrape layers of soil from the ground, and transport and spread the soil thus obtained. 3. A device for cleaning pipelines by scraping off deposits, scaling, etc.

screen 1. A device provided with openings designed to retain coarse solids in sewage (sewage screen). 2. A device for the separation of small particles from large ones in a solid waste stream; it usually takes the form of a perforated drum that rotates on an axis inclined slightly to the horizontal, or of a vibrating perforated plate. *Drum s.,* a screen in which a perforated screening medium is carried on the circumference of a revolving cylindrical drum. *Rotary s.,* a large, perforated steel cylinder that rotates slowly on a horizontal or near-horizontal axis. Solid waste fed in at one end passes along the cylinder, the rotary action spreading out the waste and allowing the smaller particles to fall through the perforations. *Vibrating s.,* a perforated or mesh plate on which material is separated by vibratory action, the smaller material passing through the perforations and the larger passing over the screen.

scrubber See *dust separator*

secator A separating device that throws mixed material on to a rotating shaft; heavy and resilient materials bounce off one side of the shaft, while light and inelastic materials land on the other and are thrown in the opposite direction. (EPA) See also *separation; separator*

secondary air See *overgrate air*

sedimentation The removal by gravity of solids carried in a liquid. (WHO)

seeding 1. The recycling of actively composting material in mechanical digestion composting processes. 2. The application of grass seed to land. Landfill sites are often seeded after topsoiling to give a tidy appearance and to reduce weed growth.

seepage The slow movement of water or gas through the pores and fissures in soil, rock, etc. without the formation of definite channels.

self-heating The rise in temperature in organic matter undergoing decomposition by the action of microorganisms.

semi-trailer A trailer forming part of an articulated vehicle.

sensor A device that can identify and code materials by pattern recognition. Infrared sensors, using reflectance spectroscopy and impact sensors, which use accelerometers, have both been used experimentally for the automatic sorting of solid waste. *Binary s.,* a sensor designed on the basis of the specific properties of one class of material. *Multiple-output s.,* a sensor that provides several different outputs from a single set of readings.

separation The systematic division of solid waste into designated categories. (EPA) *Ballistic s.,* a separation method in which a mixture of materials is dropped on to a high-speed rotary impeller; materials of different physical characteristics are then thrown off at different velocities and land in separate collecting bins. *Dry s.*, a system whereby metals of different densities are separated in a bed of iron powder or other material "fluidized" by a stream of air moving upwards through the bed. *Heavy-media s.*, separation of solid wastes into heavy and light fractions in a fluid medium whose density lies between theirs. (EPA) *Magnetic s.*, the removal of ferrous metals by means of magnets. *Source s.*, the separation of certain components of solid waste at the place of its generation, generally with the aim of facilitating the recovery of materials for recycling, e.g., wastepaper, bottles, metals. *Vibropneumatic s.*, a method, used in composting plants, of separating glass from pulverized refuse. The pulverized material is fed from a vibrating plate into an air stream that separates the lighter material from the glass.

separation plant A plant where solid waste is separated into various constituents by mechanical processes.

separator A plate or fork used in wastepaper or solid-waste baling processes to separate the material into bales while in the press. *Cinder s.*, an inclined belt used to separate cinders from refuse that has been passed through a screen to remove fine material. The screened material falls onto the belt; cinders, because of their hardness and roundness, fall off, while the organic material remains on the belt. *Inclined-belt s.*, a separating device that operates by feeding material onto an inclined belt conveyor so that heavy and resilient materials, such as glass, bounce down the conveyor, and light and inelastic materials are carried upwards by the motion of the belt. *Optical s.*, a device for separating different materials based on different reflexes to light. *Osborne s.*, a device that utilizes a pulsed, rising column of air to separate small particles of glass, metal, and other dense items from compost. (EPA) *Overband s.*, see *overband magnet*. See also *screen*

septic tank A tank, usually underground, through which sewage flows; the deposited organic matter is wholly, or partially, broken down anaerobically. (WHO) Cf. *cesspool*

service site A residential unit, commercial establishment, or other pick-up point that receives periodic solid waste collection service. (EPA)

settlement 1. The depth by which the surface of a layer of solid waste in a controlled tip sinks beneath the level when first compacted. 2. Any downward movement of a surface below a fixed horizontal reference plane. *Differential s.*, the nonuniform settlement of material from a fixed horizontal reference plane.

settling The sinking of the surface of a landfill as a result of shrinkage of dumped refuse. The degree of settling depends primarily on the kind of refuse used and the thoroughness with which it is compacted.

settling chamber 1. A chamber in the base of which suspended solids in a liquid settle for subsequent removal. 2. A chamber inserted between a furnace and its stack in which coarse particulate matter settles out of the gas stream. See also *baffle chamber*

settling velocity The velocity at which a given dust will fall out of dust-laden gas under the influence of gravity alone. (EPA)

sewage Water supply of a community after it has been fouled by various uses. It may be a combination of the liquid or water-carried domestic, municipal and industrial wastes, together with such ground-water, surface-water and storm-water as may be present. (WMO/UNESCO) *Crude s.*, sewage that has not been subjected to any treatment process.

sewer A conduit for the carriage of liquid wastes, for the most part under atmospheric pressure. (WHO)

sewerage A system for the collection and transport of sewage, including conduits, pipes and pumping stations.

shear 1. A device for cutting bulky solid waste into smaller pieces that can conveniently be passed through an incinerator plant. 2. A device for cutting scrap metal.

shredder A machine that reduces sheet material to small pieces by a tearing or cutting action. Shredders are used in solid waste treatment to reduce scrap sheet metal, automobile bodies, wastepaper and fibreboard. *Car s.*, a large hammermill pulverizing machine designed to accept the bodies and components of old cars and to shred them into scrap metal that can be recycled to the steel-making industry.

shredding The reduction of sheet material to small pieces. Sometimes also used for the pulverization of mixed solid waste in a hammermill.

shutter, dustless A shutter forming part of the dustless-loading mechanism on a refuse collection vehicle.

siftings See *riddlings*

signature The response given by an object being tested by a simple or multiple-output sensor.

silicon carbide A very hard substance that is widely used as an abrasive (e.g., in grinders) and a refractory (e.g., in furnace lining materials). It sublimes (with decomposition) at 2 700 °C.

silo A storage bunker above ground level, generally designed so that the stored material can be removed by gravity.

silt An unconsolidated sediment that results from the weathering of rock and that consists of particles whose diameters range from 1/16 mm to 1/256 mm.

sintering The fusion of particles by heat below a temperature at which they would melt, to produce a material having a cellular structure.

site specification A reference list relating to a landfill site, prepared during the project planning stage, containing information on site details, method of working, equipment required, types of waste and other relevant items.

skip or **skep** An open-topped container used for the storage and transport of solid material.

skip lift The process of lifting a skip mechanically on to a vehicle chassis designed to handle skips.

slag The nongaseous waste material formed in a metallurgical furnace; it is largely nonmetallic but usually contains some metal. Slag contains as much of the undesirable constituents of the ore as possible and is withdrawn from the furnace in the molten form. Alkaline slags retain much of the sulfur in the fuel and the ore, and the gases issuing from the furnace may be virtually sulfur-free.

slagging of refractories Destructive chemical action that forms slag on refractories subjected to high temperatures. Also a molten or viscous coating produced on refractories by ash particles. (EPA)

slave unit A vehicle that operates only at a site or plant (e.g., landfill site or transfer station) to handle trailers or containers delivered by other vehicles.

sludge 1. The accumulated solids separated from liquids, such as water or wastewater, during processing, or deposits on bottoms of streams or other bodies of water. (EAWAG) 2. The precipitate resulting from the chemical treatment, coagulation, or sedimentation of water or wastewater. (EAWAG) *Activated s.*, a flocculent sludge produced by the prolonged aeration of sewage. Due to the organisms present this sludge is a valuable means of biologically stabilizing organic matter in sewage. (WHO) *Digested s.*, sludge digested under either aerobic or anaerobic conditions until the volatile content has been reduced to the point at which the solids are relatively nonputrescible and inoffensive. (EAWAG) *Industrial s.*, the waste product, in the form of a mixture of solids and liquid, from an industrial process; also, the product of industrial wastewater treatment. *Municipal s.*, sludge obtained from the treatment of municipal wastewater. (EAWAG) *Raw s.*, settled sludge promptly removed from sedimentation tanks before decomposition has much advanced. (EAWAG) *Sewage s.*, the sludge resulting from sewage treatment processes. *Sulfate s.*, sludge produced in the desulfurization of chimney gas.

sludge cake Sludge that has been dewatered by a treatment process to a moisture content of 60–85%, depending on type of sludge and manner of treatment. (EAWAG)

sludge conditioning Pretreatment of sludge to assist its drainage and filtration. (WHO)

sludge-digestion tank See *digester*

slurry A mixture of solids and liquid in a fluid state.

smoke Suspension in the atmosphere of small particles produced by combustion. (WMO) In chemistry it is defined less restrictively as an aerosol of solid (e.g., magnesium oxide smoke) or liquid (e.g., tobacco smoke) particles arising from combustion, thermal decomposition, or evaporation. The definition of the term in air pollution legislation may vary, but is generally related to emissions that can be seen issuing from a stack or chimney; the degree of darkness of such emissions may be defined in terms of the Ringelmann scale (see *smoke chart*). See also *fume*

smoke alarm An instrument that continuously measures and records the density of smoke by determining how much light is obscured when a beam is shown through the smoke; an alarm fitted in a flue goes off when the smoke exceeds a preset density. (EPA)

smoke chart A means of assessing the darkness of a plume of smoke in terms of "smoke shade". The *Ringelmann chart* consists of four squares, each ruled into a grid of small squares, but with the thickness of the black ruling different on each. When the chart is supported some 15 m from the observer, in his line of sight to the plume, the rulings cannot be distinguished, but merge to give different shades of grey; these correspond to Ringelmann shades 1–4. The observer notes which shade provides the closest match to the plume. For shade 1, about 20% of the area of the square is occupied by the rulings; the corresponding figures for shades 2, 3 and 4 are 40%, 60% and 80% (Ringelmann shade 0 is white and shade 5 is black). This form of Ringelmann chart is little used now. The *micro-Ringelmann chart* is a reduced form on which are printed accurate reductions of Ringelmann shades 1–4, which merge to shades of grey when the card is held at arm's length. The card has a slot or hole through which the plume is viewed. The *miniaature smoke chart* is a card on which are printed four shades of grey that match Ringelmann shades 1–4 when the card is held at arm's length (strictly, 1.5 m from the eye). It is considered to be easier to use than the micro-Ringelmann chart. Normally the observed Ringelmann shade depends not only on the absorbance of the smoke, but also on its colour, the stack diameter, and the brightness of the background sky. By defining the Ringelmann number in relation to a standard sky and, if necessary, correcting for stack diameter, the Ringelmann number can be roughly correlated with the absorbance of the smoke as measured in the stack by an obscuration smokemeter.

smoke concentration In principle, the concentration of smoke particles in air, expressed in terms of mass concentration (i.e., mass of smoke particles per unit volume of air). In practice it would be extremely difficult to determine such a concentration using the definition given in the entry *smoke* (q.v.). To overcome this difficulty, any of several different conventions may be used. The amount of material collected on a smoke filter under specified conditions may be weighed and the concentration of aerosol so obtained taken as the smoke concentration. Such a procedure includes aerosol particles below about 10–20 μm in size whether they originate from the incomplete combustion of fuel or not. Another convention is to use a standard calibration curve to convert the darkness of a smoke stain into the "concentration of equivalent standard smoke" (smoke concentration). Such a curve may be obtained by running an ordinary smoke filter and a gravimetric determination of aerosol side by side. While the latter procedure is nothing more than a way of expressing the darkness of a smoke stain, or the staining capacity of the air, it has the advantage of giving the results in units that approximate more or less closely to the aerosol concentration.

smoke eye A device consisting of a light source and a photoelectric cell that measures the degree to which smoke in a flue gas obscures light. (EPA)

soil cohesion The mutual attraction between soil particles as a result of molecular forces and moisture films.

soil plasticity The property of a soil that allows it to be deformed or moulded in a moist condition without cracking or falling apart. (EPA)

soil stripping The removal of topsoil and subsoil from land.

solids *Fixed s.*, the solids remaining after ignition of a solid or evaporation of a liquid. *Total s.*, sum of dissolved and suspended solids. (WHO)

solid waste Any refuse or waste material, including semisolid sludges, produced from domestic, commercial or industrial premises or processes including mining and agricultural operations and water treatment plants.

solid waste management The purposeful, systematic control of the generation, storage, collection, transport, separation, processing, recycling, recovery and disposal of solid wastes. (EPA)

soot blower Compressed air or steam jet equipment for removing gas-borne deposits from boiler tubes.

sorting The manual separation and extraction of salvageable material from solid waste. *Binary s.*, the extraction of one class of material (e.g., ferrous metal) from inhomogeneous solid waste; the separation of the constituents of waste into two streams on the basis of physicochemical response to imposed conditions.

spalling of refractories The breaking or crushing of a refractory unit due to thermal, mechanical, or structural causes. (EPA)

spark arrester A screen-like device that keeps sparks, embers, or other ignited materials above a given size within an incinerator. (EPA)

specific volume The ratio of volume to mass (i.e., the reciprocal of mass density). The SI unit is cubic metre per kilogram, m^3/kg. The specific volume of solid wastes is usually expressed in terms of cubic metres per tonne, m^3/t (the tonne has been accepted by the Conférence générale des Poids et Mesures for general use with the SI), or — in some countries — in cubic yards per ton, yd^3/ton.

spoil Soil or rock that has been removed from its original location. (EPA)

spoil heaps Heaps of solid waste produced in mining operations.

spotter A man who guides the driver of a truck in positioning the vehicle for purposes of loading or unloading.

spray chamber A chamber equipped with water sprays that cool and clean incinerator combustion products passing through the chamber. (EPA)

spray washer See *dust separator*

spring Emergence of groundwater at the surface at a defined location. (WHO)

sprinkler system A fire protection system consisting of pipes and nozzles that automatically spray water when the ambient temperature rises to a predetermined level.

squeeze-clamp A device consisting of two horizontally opposed plates that can be fitted to a fork-lift type of truck or to a crane; used for handling baled material.

stabilization 1. The degradation of putrescible organic substances by aerobic and/or anaerobic microbic populations to yield a chemically stable product. 2. The establishment of an equilibrium between the vegetation of a locality and other environmental factors, e.g., climatic, biotic, etc.

stack A vertical passage through which products of combustion are conducted to the atmosphere. (EPA) A chimney.

stack effect The vertical movement of hot gases in a stack that results because they are hotter and therefore less dense than the atmosphere.

stack effluent Gases and suspended particles emitted from an industrial stack or chimney. (WMO) The gases are known as stack gases (or flue gases) and the term "stack solids" is frequently applied to their solid-particle content (dust, grit, etc.).

stack sampling The collecting of representative samples of gaseous and particulate matter that flows through a duct or stack. (EPA)

standard A technical specification or other document available to the public, drawn up with the consensus or general approval of all interests affected by it, based on the consolidated results of science, technology, and experience, aimed at the promotion of optimum community benefits and approved by a body recognized on the national, regional, or international level. (ECE) *Primary protection s.,* in air pollution control, an accepted maximum level of a pollutant (or its indicator) in the target, or some part thereof, or an accepted maximum intake of a pollutant or nuisance into the target under specified circumstances. *National Ambient Air Quality Standards,* a set of air quality standards for the USA issued by the Environmental Protection Agency. They are maximum permissible levels of sulfur and nitrogen oxides, carbon monoxide, hydrocarbons, photochemical oxidants, and suspended particulates and are designated as either primary or secondary standards. Primary standards are the maximum levels consistent, with an adequate safety margin, with the preservation of public health, and must be complied with within a specified time limit. Secondary standards are those judged to be necessary for protection against known or anticipated adverse effects other than health hazards (in practice they are concerned largely with effects on vegetation) and must be complied with "within a reasonable time".

star bucket See *grapple*

start-up period 1. In starting up from cold, the time required for an incinerator to reach normal combustion temperature. 2. The time that elapses between ignition of the auxiliary fuel and the release of sufficient heat from the solid waste for the combustion temperature to be maintained.

steam, superheated Water vapour at a temperature greater than 100°C, prepared by heating steam out of contact with water.

sterilization The destruction or removal, by chemical or physical (including mechanical) means, of all microorganisms in a material or on an object.

stoichiometric air See *air, theoretical*

stoker A mechanical device for feeding solid waste to a furnace. A term applied to automatic grates in general. *Step-grate s.,* an automatic furnace grate assembly with one or more "steps" or vertical drop sections intended to ensure the complete turnover of the burning material. See also *grate*

stoker drive The mechanism that actuates the sections of an automatic furnace grate.

storage pit A pit in which solid waste is held prior to processing. (EPA)

standard temperature and pressure (STP) See *normal temperature and pressure*

stream, solid waste A term applied to a single flow line in a solid waste treatment plant, e.g., a two-stream plant would have two independent treatment lines.

subsidence Settling or sinking of the land surface due to many factors such as the decomposition of organic material, consolidation, drainage, and underground failures. (EPA)

subsoil That part of the soil beneath the topsoil; usually does not have an appreciable organic matter content. (EPA)

substrate In chemistry and biology, a substance or mixture of substances on which an enzyme acts, or, in microbiology, on which a microorganism grows.

sump A pit sunk below the general level of a building or excavation to collect water or to provide storage for a pump. (WHO)

surface cracking Discontinuities that develop in the cover material at a sanitary landfill due to the surface drying or settlement of the solid waste. (These discontinuities may result in the exposure of solid waste, entrance or egress of vectors, intrusion of water, and venting of decomposition gases.) (EPA)

swill Semiliquid waste material consisting of food scraps and free liquids. (EPA)

T

tailgate A hinged door at the discharge end of a solid waste container.

tailings 1. In a separation plant, the material remaining for disposal after other material has been extracted or separated from solid waste. 2. Second-grade or waste material derived when raw material is screened or processed. (EPA)

tare weight The unladen weight of a vehicle.

thermal conduction See *heat, conduction of*

thermal conductivity The ratio between the rate of heat flow per unit area and the negative temperature gradient, or the quantity of heat that flows through unit volume per unit time per unit temperature difference between two opposing faces of the volume. The SI unit is watt per metre kelvin, $W \cdot m^{-1} \cdot K^{-1}$ or $W/(m \cdot K)$, but the degree Celsius (°C) may be used in place of the kelvin. Symbol: λ

thermal efficiency The ratio of heat used to total heat generated. (EPA)

thermal shock The stress to which a body or material (e.g., the refractory lining of a furnace) is subjected as a result of sudden and large temperature changes.

thermal shock resistance The ability of a material to withstand sudden heating or cooling or both without cracking or spalling. (EPA)

thixotropy The property exhibited by certain substances liquefying when subjected to a strain, such as agitation, and setting or hardening again when the strain is removed. It may also be defined as a property whereby a gel is reversibly converted into a sol. Certain types of clay are thixotropic.

tidal marsh Low flat marshlands traversed by interlaced channels and tidal sloughs and subject to tidal inundation; normally, the only vegetation present is salt-tolerant bushes and grasses. (EPA)

tip An area of land or a hole in the ground used for the ultimate disposal of solid waste. *Controlled tip,* see *sanitary landfill*

tipper, hydraulic A device that unloads a transfer trailer by raising its front end to a 70 degree angle. (EPA)

tipping beam A timber, concrete or steel beam at the edge of a solid waste reception bunker; it acts as a stop for vehicles backing on to the bunker.

tipping, controlled See *sanitary landfilling*

tipping floor Unloading area for vehicles that are delivering solid waste to an incinerator or other processing plant. (EPA)

toe The bottom of the working face at a sanitary landfill. (EPA)

topsoil The topmost layer of soil; usually refers to soil that contains humus and is capable of supporting good plant growth. (EPA)

tote barrel See *container*

totting Unauthorized scavenging.

trace element 1. An element present in very low concentrations (the "trace" level is defined as having an upper limit of $100\,\mu g/g$; a lower limit is at present under study by the International Union of Pure and Applied Chemistry). Concentrations are quoted in terms of the element itself, although trace elements are almost always present in the air in the form of oxides or other compounds. Common examples of trace elements are lead, copper, zinc, arsenic and vanadium. 2. In biology, an element that, while essential for the nutrition of an organism, is required only in minute amounts; an excess of such an element may be toxic.

tramp material Items in an assortment of material being processed, e.g., solid waste, that are detrimental to the processing machinery or to the final product.

transfer haulage cost The total transport costs involved in the haulage of solid waste from a transfer station to a disposal facility, excluding transfer station costs.

transfer station A place where refuse is assembled before being transported to disposal sites or treatment stations. (EAWAG)

transfer system *Compaction pit t.s.*, a transfer system in which solid waste is compacted in a storage pit by a crawler tractor before being pushed into an open-top transfer trailer. (EPA) *Direct-dump t.s.* or *direct-loading t.s.*, a system by which solid waste is unloaded direct from a collection vehicle into an open-top transfer trailer or container. *Internal compaction t.s.*, a transfer method in which the reciprocating action of a hydraulically powered bulkhead contained within an enclosed trailer packs solid waste against the rear doors. (EPA)

transfer vehicle A large-capacity vehicle used to carry solid waste from a transfer station to a disposal site.

transpiration Loss of water by plants through the surface of their leaves. (WHO)

transportation *Bulk t.*, the movement of solid waste in a large-capacity vehicle from a transfer station to a landfill site or treatment plant. *Hydraulic t.*, the movement of solid waste by a carrier liquid in pipelines. *Pneumatic t.*, the movement of solid waste through pipelines by means of an air stream.

trash Dry solid waste generated at domestic premises, offices, etc. (e.g., paper, cans, bottles) as opposed to garbage (USA usage).

treatment, biological The treatment of water or sewage for the removal of organic matter with the assistance of living organisms.

treatment plant A plant where waste is processed to alter its physical or chemical characteristics in order to facilitate disposal, e.g., incineration plant, composting plant, pulverization plant.

trimming The process of levelling off a load of solid waste in an open-topped vehicle or barge.

truck scales An apparatus for weighing vehicles; a *weighbridge* (q.v.) or platform scale.

turning circle The curve of minimum radius required by any given vehicle to turn in a forward direction. Turning circles may be given as "between kerbs" or "between walls", the latter allowing for the overhang of parts of the vehicle.

tuyeres Opening or ports in a grate through which air can be directed to improve combustion. (EPA)

tynes Steel teeth on a grab or mechanical bucket that dig into the material to be lifted.

U

U-blade A dozer blade with an extension on each side; these protrude forwards at an obtuse angle to the blade and enable it to handle a larger volume of solid waste than an ordinary blade.

underfire air The forced or induced air that is supplied in controlled quantity and direction beneath a grate and passes through a fuel bed. See also *combustion air*

unloading bulkhead A steel plate that ejects waste out the rear doors of an enclosed transfer trailer. It is propelled by a telescoping, hydraulically powered cylinder that traverses the length of the trailer. (EPA)

utility, private A private business [in the USA] that collects, processes, and disposes of solid waste under a government licence or monopoly franchise. (EPA)

utilization factor A measure of plant or machine performance in terms of actual use time compared with total availability time.

V

valve chamber A chamber in a pneumatic conveying system housing a shut-off valve at the junction of a vertical chute and a lead-in pipe to the horizontal transporter pipes.

vector Strictly, any animal that carries an infectious agent from an infected host (or its waste products) to another host or its immediate environment. In practice, the term is restricted to an animal belonging to a phylum that is different from the one to which the host belongs, and is applied most frequently to arthropods.

vehicle, moving-floor A vehicle fitted with a floor that consists of transverse hinged steel plates attached to chains; it can be moved backwards or forwards mechanically.

vehicle, satellite A small collection vehicle that transfers its loads into a larger vehicle operating in conjunction with it. (EPA)

vehicle weight, gross The combined weight of a vehicle and its load.

vehicle weight, permissible gross The maximum laden weight at which a vehicle can legally be used on public roads.

venturi scrubber See *dust separator*

vertical cell system See *vertical digester* under *digester*

vitrification The process whereby high temperatures effect permanent chemical and physical changes in a ceramic body, most of which is transformed into glass. (EPA)

volatile matter Apparent loss of matter from a residue ignited at $600 \pm 25\,°C$ for a period of time sufficient to reach a constant weight of residue, usually 10–12 min. (EAWAG)

volatile solids The material lost from a dry solid waste sample that is heated until it is red in an open crucible in a ventilated furnace. The weight of the volatile solids is equal to that of the volatile matter plus that of the fixed carbon. (EPA)

volume reduction ratio The ratio of input to output volume in a solid waste treatment plant.

W

wall *Air-cooled w.*, a refractory wall in an incinerator, containing internal passages through which cooling air is forced or flows by convection. *Battery w.* or *division w.*, a double or common wall between two incinerator combustion chambers; both faces are exposed to heat. *Bridge w.*, a partition between chambers over which the products of combustion pass. *Core w.*, in a battery wall, those centre courses of brick none of which is exposed on either side. *Curtain w.*, a refractory construction or baffle that deflects combustion gases downwards. *Gravity w.*, a furnace wall supported directly by the foundation or floor of a structure. (EPA) *Insulated w.*, a furnace wall in which refractory material is installed over insulation. *Sectionally-supported w.*, a furnace or boiler wall that consists of special refractory blocks or shapes mounted on and supported at different heights by metal hangers. *Unit suspended w.*, a furnace wall or panel hung from a steel structure. *Water w.*, the interior lining of a furnace, constructed from metal tubes in which water circulates; the term is used in connexion with heat recovery when solid waste is incinerated.

waste Materials that are not prime products, the principal source of which is the manufacturing industries. Wastes arise from such industries both as the by-products of manufacturing processes and as consumer discards of manufactured products. See also *refuse; solid waste*. *Agricultural w.*, waste arising from agricultural or farming activities; it includes animal waste. *Bulky w.*, a special category of waste which, owing to its bulky character, cannot be stored or placed in a dustbin or refuse sack. *Commercial w.*, see *trade waste*. *Consumer w.*, manufactured products that have reached the end of their useful life or have become obsolete or unwanted and are discarded. *Container w.*, a term employed in the wastepaper trade for used corrugated and solid fibreboard cases free from board of any description. *Demolition w.*, rubble and other waste material arising from the demolition or reconstruction of buildings. *Domestic w.*, solid waste arising from a residential environment. *Farm w.*, see *agricultural waste*. *Food w.*, animal and vegetable waste resulting from the handling, storage, sale, preparation, cooking, and serving of foods; in the USA, commonly called garbage. (EPA) *Food-processing w.*, waste resulting from operations that alter the form or composition of agricultural products for marketing purposes. (EPA) *Garden w.*, plant clippings and other discarded material from gardens. *Hazardous w.*, waste that, because of its physical, chemical or biological characteristics, requires special handling and disposal procedures to avoid risk to health or damage to property. *Incombustible w.*, material that cannot be burned at normal solid waste incinerator temperatures (800-1 000 °C); mainly inorganic materials such as metals, glass and ceramics. *Industrial w.*, waste originating from industrial activities. (EAWAG) *Inert w.*, waste that retains its form and characteristics indefinitely and, apart from its physical presence, has no perceptible effect on the environment. *Institutional w.*, solid waste originating from educational, health care, research and similar establishments. *Medical w.*, waste material, such as surgical dressings, from nursing homes or domestic premises where patients are under nursing care. Such wastes are generally collected separately and require special handling and disposal. *Organic w.*, solid waste consisting mainly or entirely of organic matter. *Packing-station w.*, waste from poultry-processing plants, consisting mainly of feathers, inedible offal and blood. *Pesticide w.*, the residue arising from the manufacture, handling or use of pesticides. *Process w.*, waste arising from the processing of material in a manufacturing industry. *Radioactive w.*, waste consisting of or containing radioactive materials, requiring special handling and disposal procedures. *Residential w.*, see *domestic waste*. *Special w.*, waste consisting of or containing hazardous materials, either solid or liquid, such as flammable, explosive, toxic, radioactive or pathogenic materials. *Terminal w.*, waste in which pollutants are "fixed" and better controlled. In this sense, most solid wastes are terminal wastes. *Trade w.*, solid waste arising in shops, offices and other commercial premises that do not produce manufactured products. *Vegetation w.*, wastes arising from crop production and other agricultural or horticultural activities. *Woodpulp w.*, wood or paper fibre residue resulting from a manufacturing process. (EPA)

waste disposal The orderly process of discarding useless or unwanted material.

waste paper Newspapers, magazines, cartons and other paper separated from solid waste for the purpose of recycling.

waste sources Agricultural, residential, commercial, and industrial activities that generate wastes. (EPA)

wastewater See *sewage*

water *Boiler w.*, the water present in a boiler when steaming is, or has been, taking place. *Capillary w.*, underground water that is held above the water table by capillary action. *Make-up w.*, water added to the boiler-feed system to make up for losses. *Surface w.*, water on, or flowing over, the surface of land.

water conditioning The treatment of water; more especially, the treatment of boiler feed-water.

water table The free surface of the zone of saturation, i.e., the surface separating the upper unsaturated from the lower saturated soil.

water transport 1. Transport by means of barges, ships, etc. 2. Transport of solids in a water stream. (Both EAWAG)

weighbridge A platform machine for weighing vehicles.

wet line kit A system used in conjunction with an enclosed transfer trailer to power its unloading bulkhead. The bulkhead's hydraulic pump is driven by a power-take-off unit on the semi-tractor's transmission. (EPA)

white goods Discarded kitchen and other large, enamelled appliances. (EPA)

windbox A chamber below a furnace grate or surrounding a burner, through which air is supplied under pressure to burn the fuel. (EPA)

windrowing, modified *Composting* (q.v.) in windrows in which controlled amounts of air are blown through the material to speed up fermentation.

working face That portion of a sanitary landfill where waste is discharged by collection trucks and is compacted prior to placement of cover material. (EPA)

X, Y

yard tractor A small semi-tractor used exclusively for manoeuvering transfer trailers into and out of loading position. (EPA)

yield The amount of solid waste generated per unit of population per unit time.

Z

zig-zag air classifier A static device for the separation of mixed waste in a vertical duct that repeatedly changes direction.

zone, drying The section of the furnace where the solid waste is partially dried by hot gases from the combustion of material in the burning zone of the grate, and by heat stored in refractory furnace linings.

zone of aeration The area above a water-table where the interstices (pores) are not completely filled with water. (EPA)

zone of capillarity The area above a water-table where some or all of the interstices (pores) are filled with water that is held by capillarity. (EPA) See also *capillary water*

zoning The allocation of areas of land in urban development plans for particular classes of use, e.g., industrial, housing, etc.

zymogenic Causing or pertaining to fermentation; as *zymogenic bacteria*, bacteria that cause fermentation.

SOURCES OF DEFINITIONS

BSI   BRITISH STANDARDS INSTITUTION

The storage and on-site treatment of refuse from buildings. Part 1. Residential buildings. London, 1972 (Code of Practice CP 306: part 1). (Material from this standard has been reproduced by permission of the British Standards Institution, 2 Park Street, London W1A 2BS from whom complete copies can be obtained.)

EAWAG   WHO INTERNATIONAL REFERENCE CENTRE FOR WASTES DISPOSAL

Solid wastes thesaurus. Dübendorf, Institute for Water Resources and Water Pollution Control, Swiss Federal Institutes of Technology, 1973.

ECE   UNITED NATIONS ECONOMIC COMMISSION FOR EUROPE

Recommendations on standardization policies. Geneva, 1977 (document ECE/STAND/17). (The glossary contained in this document has been reprinted by the International Organization for Standardization as ISO Guide 2.)

EPA   US ENVIRONMENTAL PROTECTION AGENCY

Solid waste management glossary. Washington, DC, Federal Solid Waste Management Program, 1972 (Publication SW-108ts).

FRG   FEDERAL REPUBLIC OF GERMANY

Federal Law on Protection against Emissions, 15 March 1974.

ISO   INTERNATIONAL ORGANIZATION FOR STANDARDIZATION

ISO, 1   Vocabulary of terms relating to solid mineral fuels, Part I. Terms relating to coal preparation. Geneva, 1970 (ISO/R 1213).

ISO, 2   Cleaning equipment for air and other gases: vocabulary. Geneva, 1975 (Draft International Standard ISO/DIS 3649).

POLLOCK   POLLOCK, K.M.

Compactor handbook: the use, selection and economics of refuse compaction. New York, Communication Channels, 1973 (copyright © 1973 by *Solid wastes management* magazine).

SCOTT   SCOTT, J.S.

A dictionary of civil engineering. Harmondsworth, Penguin Books, 1958 (copyright © 1958 by John S. Scott).

UK   UNITED KINGDOM

Alkali Act, Works Order 1963, Chemical Incineration Works.

WHO   WHO REGIONAL OFFICE FOR EUROPE

A glossary of water and sewage terms used in sanitary engineering practice. Copenhagen 1956 (EUR/San.Eng(1956)/8 (Pt I)).

WMO   WORLD METEOROLOGICAL ORGANIZATION

International meteorological vocabulary. Geneva, 1966 (WMO/OMM/BMO – No. 182.TP.91).

WMO/UNESCO   WORLD METEOROLOGICAL ORGANIZATION & UNITED NATIONS EDUCATIONAL, SCIENTIFIC AND CULTURAL ORGANIZATION

International glossary of hydrology. Geneva, 1974 (WMO/OMM/BMO – No. 385).